普通高等教育机械类专业教材

Introduction of Mechanical Engineering

曹蕾蕾　宋绪丁　叶　敏　主　编

人民交通出版社股份有限公司
北　京

Abstract

This book is an elementary mechanical engineering textbook that is fundamental, comprehensive, and forward-looking. The book contains six chapters. Chapter 1 introduces overview and history of mechanical engineering; Chapter 2 presents machine design theory; Chapter 3 elaborates typical mechanical parts; Chapter 4 introduces mechanical manufacturing technology; Chapter 5 demonstrates advanced manufacturing technology; Chapter 6 concludes modern design methods and the applications.

This book can serve as an introductory textbook for international students in China specializing in mechanical engineering and teaching or reference material for students studying in other related fields to acquire mechanical engineering knowledge.

图书在版编目(CIP)数据

机械工程导论 = Introduction of Mechanical Engineering:英文/曹蕾蕾,宋绪丁,叶敏主编. —北京:人民交通出版社股份有限公司, 2023.12

ISBN 978-7-114-19120-6

Ⅰ.①机… Ⅱ.①曹… ②宋… ③叶… Ⅲ.①机械工程—英文 Ⅳ.①TH

中国国家版本馆 CIP 数据核字(2023)第 231685 号

书　　名: **Introduction of Mechanical Engineering**
著 作 者: 曹蕾蕾　宋绪丁　叶　敏
责任编辑: 钟　伟　李佳蔚
责任校对: 赵媛媛　龙　雪
责任印制: 刘高彤
出版发行: 人民交通出版社股份有限公司
地　　址: (100011)北京市朝阳区安定门外外馆斜街 3 号
网　　址: http://www.ccpcl.com.cn
销售电话: (010)59757973
总 经 销: 人民交通出版社股份有限公司发行部
经　　销: 各地新华书店
印　　刷: 北京虎彩文化传播有限公司
开　　本: 787 × 1092　1/16
印　　张: 9.25
字　　数: 278 千
版　　次: 2023 年 12 月　第 1 版
印　　次: 2023 年 12 月　第 1 次印刷
书　　号: ISBN 978-7-114-19120-6
定　　价: 39.00 元

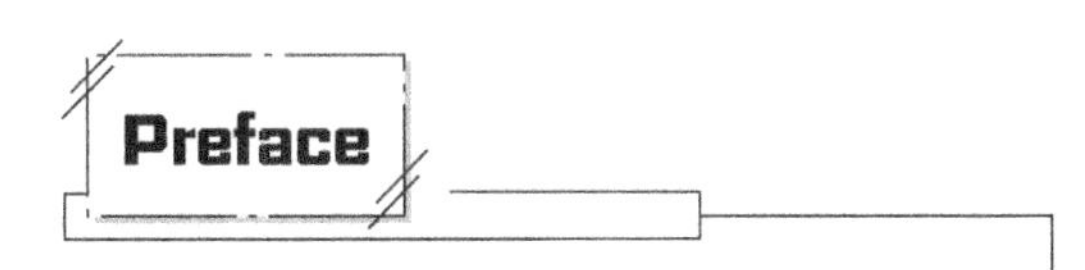

Mechanical engineering, as a comparatively wide and complex discipline, covers knowledge from design, manufacturing, automation to materials and other aspects. In recent years, more and more international students have chosen to study in China, and are full of interest and enthusiasm in mechanical engineering. However, with the problem of language and culture differences, many students always encounter some difficulties at the beginning of study. Therefore, we hope to provide them with a comprehensive overview of the mechanical engineering discipline through this textbook " Introduction to Mechanical Engineering" to help them lay a solid disciplinary foundation on mechanical engineering.

This textbook is concise, easy to understand, comprehensive, and systematic. It introduces the basic knowledge of mechanical engineering, including its definition, historical development, and overview of the related fields, such as typical mechanical components, traditional mechanical manufacturing technology, advanced manufacturing technology, modern design methods, etc. The main features of the book are as follows:

(1) Full English writing is suitable for international students in China.

Considering the diversity and globalization background of international students in China, this book is written in full English to meet their learning needs. At the same time, it can also help Chinese students learn about the English expressions of terminology, concepts, and theories in mechanical engineering. Furthermore, it can also motivate the students to have a deep understanding of professional literature and cutting-edge research in the world, and expand their international academic and career opportunities.

(2) Emphasize basic knowledge and combine it with engineering applications.

This book focuses on introducing the basic knowledge of mechanical engineering, starting from principles and theories, then systematically addresses the basic concepts and fundamentals of mechanical engineering. At the same time, it not only stays at theoretical level, but also combines theorical knowledge with practical engineering applications which can stimulate students' learning interests and help them understand the connotations of knowledge points.

(3) Involve modern mechanical engineering technology.

Mechanical engineering is developing rapidly and new technologies are emerging constantly. Therefore, this book specifically involves content on modern mechanical engineering technology, including advanced design technology, advanced manufacturing technology, virtual technology, etc., which garantees the book forward-looking and cutting-edge, and helps students grasp the latest trends in the industry and master the latest technological applications.

Participants in the compilation of this textbook include Cao Leilei (Chapter 2, Chapter 3, Chapter 5), Ye Min (Chapter 1), Song Xuding (Chapter 4), and Bai Jie (Chapter 6). The entire book is edited by Cao Leilei, Song Xuding, and Ye Min.

The compilation of this textbook is supported by the Chang'an University Textbook Fund, and we would like to express our special gratitude.

During the compilation process, the editor drew on and referred to some content from other books, journals, and materials both domestically and internationally. We would like to express our sincere gratitude to the relevant authors! Special thanks also for the editors and reviewers of China Communications Press Co., Ltd, whose guidance and assistance are indispensable for the publication of this book.

Due to limitations in professional skills and subject knowledge, it is inevitable that this book may be inappropriate. We kindly request readers of this book to criticize and provide valuable feedback.

Authors
August 2023

Contents

Chapter 1 Overview and History of Mechanical Engineering

1.1 Definition of Mechanical Engineering

Mechanical engineering is the discipline that applies engineering, physics, engineering mathematics, and materials science principles to design, analyze, manufacture, and maintain mechanical systems. It is one of the oldest and broadest engineering disciplines.

Mechanical engineering requires an understanding of core areas, including mechanics, dynamics, thermodynamics, materials science, structural analysis, and electricity. In addition to these core principles, mechanical engineers use tools such as Computer-Aided Design (CAD) and Computer-Aided Manufacturing (CAM) to design and analyze manufacturing plants, industrial equipment and machines, heating and cooling systems, transport systems, aircraft, watercraft, robotics, medical devices, weapons, and others. It is the branch of engineering that involves the design, production, and operation of machinery.

1.2 History of Machinery Development

The history of mechanical engineering is a rich and diverse journey that spans thousands of years. Mechanical engineering can be traced back to ancient civilizations and has evolved through various technological advancements to become the multifaceted discipline it is today. Here is an overview of the key historical milestones in mechanical engineering.

1.2.1 Ancient Machinery

The ancient machinery, although rudimentary compared to modern technology, possessed several characteristics and technologies that were innovative and influential for their time. Some of the critical characteristics and technologies of ancient machinery include:

(1) Simplicity and Mechanical Advantage. Ancient machinery was often simple in design and construction, utilizing basic mechanical principles to achieve specific tasks. Inventors leveraged concepts like levers, pulleys, and inclined planes to create machines with significant mechanical advantages, allowing them to move heavy objects or perform labor-intensive tasks more efficiently.

(2) Reliance on human or animal power. Most ancient machines were human-powered or animal-powered. Humans or animals provided the motive force needed to operate these machines, making them heavily dependent on manual labor.

(3) Water-powered machinery. In many ancient civilizations, water was harnessed to power various machines. Water wheels and watermills were used for grinding grain, sawing wood, and pumping water. Water power was a crucial energy source that allowed for increased productivity and automation of certain processes.

(4) Military Engineering. Ancient machinery played a significant role in military applications. Siege engines, such as trebuchets, ballistae, and battering rams, were developed for attacking fortified structures.

(5) Mechanical Clocks and Sundials. Ancient civilizations devised methods for measuring time using mechanical clocks and sundials. Water clocks was the earliest timekeeping devices, relying on the steady flow of water to indicate the passage of time.

(6) Gearing Systems. Ancient engineers used rudimentary gear mechanisms, such as spur and bevel gears, in various applications. Gears were employed in mills, water wheels, and astronomical instruments, demonstrating early knowledge of gear ratios and motion transmission .

(7) Construction and Architecture. Ancient machinery was vital in construction and architecture. Engineers utilized cranes, ramps, and hoists to lift and move heavy stones and building materials while constructing temples, pyramids, and other monumental structures.

(8) Metallurgy. The development of metallurgy allowed ancient civilizations to create tools and machines with improved strength and durability. Techniques like smelting and alloying metals were employed to produce materials suitable for various mechanical applications.

(9) Astronomy and Navigation. Ancient astronomers developed instruments and mechanical devices for tracking celestial bodies and aiding navigation. They used devices like astrolabes and quadrants for celestial observations and positioning.

These ancient civilizations' engineering achievements showcased their understanding of mechanical principles and innovative problem-solving abilities. While the ancient mechanical engineering techniques might seem rudimentary compared to today's technology, they formed the foundation upon which modern mechanical engineering has been built, and many of their inventions and discoveries continue to influence engineering practices today.

1.2.2 Modern Machinery

Modern machinery has seen significant advancements and innovations compared to ancient machinery. The characteristics and technologies of modern machinery include.

(1) Automation and Robotics. Modern machinery often incorporates automation and robotics, reducing the need for direct human involvement in repetitive or dangerous tasks. Robotic systems can perform precise and complex operations in manufacturing, assembly, and

logistics, improving efficiency and safety.

(2) Advanced Materials. Modern machinery benefits from using advanced materials, including high-strength alloys, composites, and nanomaterials. These materials offer improved performance, reduced weight, and enhanced durability, making machines more efficient and capable of handling higher loads.

(3) Computer-Aided Design and Simulation. Computer-Aided Design (CAD) software allows engineers to create detailed 3D machinery models before fabrication. Simulation tools enable engineers to test and optimize designs virtually, reducing development time and costs while ensuring higher reliability and performance.

(4) Precision Manufacturing. Modern machinery uses precision manufacturing techniques, such as Computer Numerical Control (CNC) machining and additive manufacturing (eg. 3D printing). These technologies enable the production of complex components with tight tolerances and high repeatability.

(5) Digital Connectivity and IoT. The Internet of Things (IoT) has enabled the integration of sensors and connectivity in modern machinery. This connectivity allows machines to gather and exchange data, enabling real-time monitoring, predictive maintenance, and remote control.

(6) Energy Efficiency. Modern machinery places a strong emphasis on energy efficiency and sustainability. Engineers design machines to optimize energy consumption, reduce emissions, and use renewable energy sources when possible.

(7) Smart Systems and AI. Smart systems and Artificial Intelligence (AI) are increasingly incorporated into modern machinery. AI algorithms can optimize machine operations, detect anomalies, and enable machines to adapt to changing conditions autonomously.

(8) Integrated Control Systems. Modern machinery often employs sophisticated control systems, such as Programmable Logic Controllers (PLCs) and Supervisory Control and Data Acquisition (SCADA) systems. These systems provide centralized control and monitoring of complex processes.

(9) Miniaturization and Portability. Advancements in microelectronics and miniaturization have led to the development of compact and portable machinery. Devices such as smartphones, laptops, and wearable technology are examples of modern machinery that offer high functionality in small form factors.

(10) Cutting-Edge Sensors. Modern machinery utilizes a wide range of sensors for data acquisition and feedback. These sensors provide information about temperature, pressure, vibration, position, and other parameters, enabling precise control and condition monitoring.

(11) Human-Machine Interfaces (HMI). User-friendly HMIs are essential in modern machinery, allowing operators to easily interact with and control complex systems. Touchscreens, graphical interfaces, and voice commands are common features in modern machinery interfaces.

(12) Integration with Cloud Computing. Modern machinery leverages cloud computing for data storage, analytics, and collaboration. Cloud-based solutions facilitate seamless sharing of data and access to computational resources from anywhere in the world.

The characteristics and technologies of modern machinery are constantly evolving as advancements in science and technology continue to shape the engineering landscape. These innovations have increased productivity, improved safety, and enhanced functionality, revolutionizing industries and our daily lives.

1.3 Fields of Mechanical Engineering

The mechanical engineering discipline is the science of the theory, method, and technology of mechanical systems and products' performance, design, and manufacture, including two major fields of mechanics and manufacturing science.

Mechanics is a science that studies mechanical structure and system performance and its design theory and method, including the mechanism, transmission, dynamics, strength, tribology, design, bionics, micromechanics, and interface mechanics involved in the manufacturing process and the mechanical system.

Manufacturing science is the science of manufacturing processes and systems. It covers product design, forming and manufacturing (casting, plastic forming, joint forming, mold manufacturing, surface engineering, etc.), processing and manufacturing (ultra-precision machining, efficient processing, non-traditional machining, complex surface processing, measurement and instrument, equipment design and manufacturing, surface functional structure manufacturing, micromanufacturing, nanomanufacturing, biomimetic and raw material production), and operation and management of the manufacturing system and other science.

There are four classical second-class disciplines belong to the first-class discipline of Mechanical Engineering: mechanical manufacture and automation, mechanical design and theory, mechatronics engineering, and vehicle engineering.

1.3.1 Mechanical Manufacture and Automation

Mechanical manufacturing and automation are one of the engineering disciplines studying mechanical manufacturing theory, manufacturing technology, automated manufacturing systems, and advanced manufacturing modes. The discipline combines the latest developments in related fields, giving manufacturing technology, systems, and models a new look. The goal of mechanical manufacture and automation is clear, it combines mechanical equipment and automation through the way of computers to form a series of advanced manufacturing technologies, including CAD (computer-aided design), CAM (computer-aided manufacturing), and FMS (flexible manufacturing system) and so on. Finally, a large-scale computer-integrated manufacturing system (CIMS) is formed, making the traditional machining process qualitatively leap. Specific applications in the industry include computer numerical control (CNC) machine tools, machining centers, and so on.

1.3.2 Mechanical Design and Theory

Mechanical design and theory is a fundamental technical discipline to analyze, synthesize, quantitatively describe, and control the performance of machinery. It is a brief introduction to the detailed work processes and procedures of mechanical engineering. Therefore, work principles, motion and dynamic properties, strength and life, vibration and noise, friction, wear and lubrication, mechanical innovation and design, and modern design calculation methods of various machines, mechanisms, and parts are mainly studied.

1.3.3 Mechatronic Engineering

Mechanical and electronic engineering, commonly known as mechatronics, is a kind of mechanical engineering and automation. Mechanical and electronic engineering majors include basic theoretical knowledge and mechanical design and manufacturing methods, the application ability of computer software and hardware, and the ability to design, manufacture, test, and develop various kinds of mechanical and electrical products and systems. Mechanical and electronic engineering is the product of the high-speed development of science and technology and the interlinking of disciplines. It breaks the traditional classification of disciplines and integrates many technical characteristics. It represents the emergence of new technologies, new ideas, new research methods, and new research objectives.

1.3.4 Vehicle Engineering

Vehicles are widely used in modern society. They are related to the revitalization and development of the automobile and transportation industry, one of the pillar industries of China's economic construction, and have a great influence on the modernization of agriculture and the modernization of national defense equipment. From the early stages of vehicle engineering, it involves mechanics, mechanical design, material, fluid mechanics, and chemical industry, and extends to the interpenetrating and interrelated subjects such as mechatronic engineering, mechanical design and theory, computer, electronic technology, measurement technology, control technology, and other disciplines, and further touches on broad fields such as medicine, physiology, and psychology, forming a comprehensive discipline and engineering technology covering a variety of new and high technologies.

According to the characteristics of the industry, this field covers the design and manufacture of automobiles and tractors, military vehicles, locomotive vehicles, engineering vehicles, energy power, and so on.

Chapter 2 Machine Design Theory

2.1 Main Contents of Mechanical Design

2.1.1 Concept of Mechanical Design

Mechanical design is a working process that conceives, analyzes, and calculates the working principle, structure, motion mode, transmission mode of force and energy, material, shape, and size of each part, lubrication method, etc., of the machine according to the user's requirements, and converts them into specific descriptions as the basis for manufacturing.

Mechanical design is an essential part of mechanical engineering, the first step of mechanical production, and the most crucial factor determining mechanical performance. The goal of mechanical design is to design the best machine under various limited conditions (such as materials, processing capacity, theoretical knowledge, and computing means), that is, to make the optimal design. Optimization design needs to comprehensively consider many requirements, generally including the best working performance, the lowest manufacturing cost, the smallest size and weight, the highest reliability in use, the lowest consumption, and the least environmental pollution. These requirements are often contradictory, and their relative importance varies with the type and use of machinery. In the past, design optimization mainly depended on the designers' knowledge, experience, and vision. With the development of the basic mechanical engineering theory, system analysis, and other new disciplines, the accumulation of technical and economic data for manufacturing, and the popularization and application of computers, optimization has gradually abandoned subjective judgment and relied on scientific calculation.

The design of industrial machinery, especially the mechanical design of the whole system, must depend on the relevant industrial technologies. Therefore, specialized mechanical design disciplines, such as agricultural machinery design, mining machinery design, pump design, compressor design, steam turbine design, internal combustion engine design, machine tool design, and so on, have emerged. However, these professional designs have many common technologies.

2.1.2 Classification of Mechanical Design

Mechanical design can be divided into new design, inheritance design, and variant design.

(1) New design.

Apply mature science and technology to design new machinery that has not been used.

(2) Inheritance and design.

According to the users' experience and technical development, the existing machinery shall be designed and updated to improve performance and reduce manufacturing and operating costs.

(3) Variant design.

Some modifications or additions and subtractions are performed to the existing machinery to develop variant products different from the standard machinery in order to fulfill the new requirements.

2.1.3 Conventional Mechanical Design Method

Mechanical design has been practiced for hundreds of years, forming some conventional mechanical design methods, which can be summarized as follows.

1) Theoretical design method

Theoretical design is based on the design theory and experimental data. It can be divided into two parts: design calculation and check calculation.

(1) Design calculation.

Calculate the size and shape of parts to meet the requirements of motion, load conditions, and material properties of parts.

(2) Check calculation.

Check calculation shall be done by analogy method, experimental method, etc. For instance, once the size and shape of the parts are calculated, they should be checked by theoretical formulae, such as the stress analysis method (when strength is the design criterion) or the deformation analysis method (when stiffness is the design criterion).

The theoretical design method can generally obtain accurate and reliable results, and the most critical parts are designed by this method.

2) Empirical design method

Empirical design is a design conducted based on the designer's working experience or the users' experience. It can apply to some minor parts or parts that are not mature in theory.

The empirical design method is effective for those parts with typical structural shapes. For example, the empirical design method can be adopted for the design of box-type, frame-type, and transmission parts.

3) Model experiment design method

Model experiment design is a method for the important parts with huge sizes and complex structures, especially some heavy parts. First, make small-size prototypes for the parts or machines, then test their various characteristics through experiments, and finally, modify and improve the design according to the experimental results.

The model experiment design method is time-consuming and expensive, so it is generally applicable to particularly significant mechanical designs only, such as new and heavy equipment, aircraft fuselage, new ship hull, etc.

2.2 Mechanical Design Process

2.2.1 Main Procedures of Mechanical Design

(1) Determine design tasks.

Determine design tasks according to user orders, market needs, and new scientific research achievements.

(2) Preliminary design.

Preliminary design includes determining the working principle and basic structure of machinery, conducting motion design and structural design, and drawing preliminary sketches.

(3) Technical design.

Technical design includes design modification (according to preliminary review comments), drawing of all parts, and a second review.

(4) Working drawing design.

Working drawing design includes the final modification (according to the comments of the second review), the preparation of all working drawings (such as part drawings, component assembly drawings, and general assembly drawings), and the preparation of all technical documents (such as part lists, wearing parts lists, instructions for use, etc.).

(5) Type design.

Type designs are used for mass production. Preliminary design procedures can be omitted for simple design tasks (such as the design of simple machinery, inheritance design or variant design of general machinery, etc.).

2.2.2 Mechanical Design Stage

The quality of a machine depends on the design quality. The effect of the manufacturing process on machine quality is essential to achieve the quality specified in the design. Therefore, the design stage of the machine is the key to the machine.

Only by combining inheritance with innovation can we design high-quality machines. A scientific design procedure must be established in order to improve the design quality. Although it is impossible to list a unique program that is effective in any case, the design program of a machine can be shown in Figure 2-1.

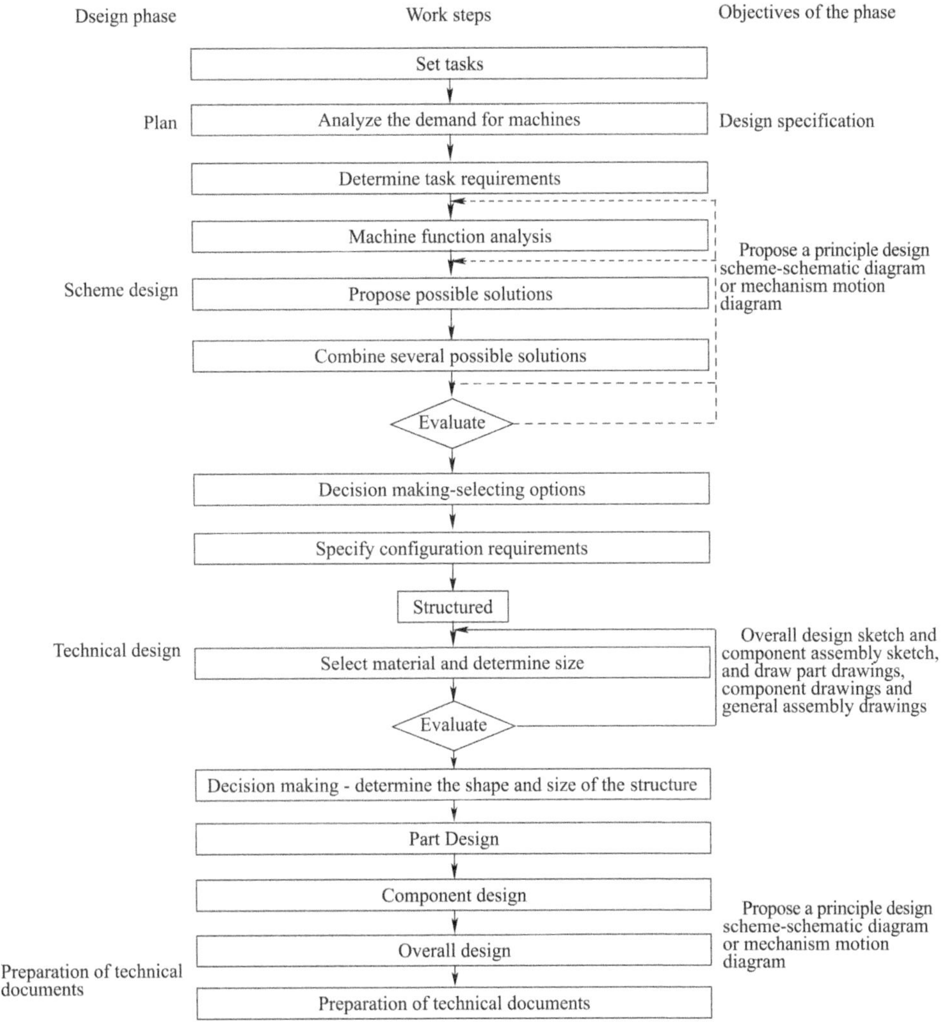

Figure 2-1 General procedure of machine design

A brief introduction of each stage is provided below.

1) Plan

In the planning stage, the demand for the designed machine should be fully investigated and analyzed. Through analysis, the functions that the machine should be further clarified, and the constraints determined by environment, economy, processing, time limit, and other aspects are proposed for future decisions. Therefore, the overall requirements and details of the design task shall be written clearly, and the design task book shall be formed as the summary of this stage. The design specification shall generally include the estimation of the machine's function, economy and environmental protection, manufacturing requirements, basic user requirements, and the deadline for the task. These requirements should be given by a reasonable range, not an accurate number.

2) Scheme design

Specific schemes of various actuators can be formulated according to different working principles. Taking thread cutting as an example, it is possible to rotate the workpiece while making the tool move in a straight line, such as cutting the thread on an ordinary lathe, or it is possible to rotate the tool while fixing the workpiece, such as machining the thread with a die. That means several different structural schemes may for the same working principle.

Of course, there are many options for the prime mover. Because of the universality of electric power supply and the development of electric drive technology, motors are mainly used for most fixed machinery. Thermal prime movers are mainly used for transport, construction machinery, or agricultural machinery. Even for the motor, there are also choices of AC and DC, high speed and low speed, etc.

The transmission part of the scheme is more complex and diverse. It can be designed by multiple mechanisms or by combinations of different mechanisms. Among the different schemes, only a few are technically feasible. These feasible schemes should be comprehensively evaluated from the technical, economic, and environmental aspects. Taking the economic evaluation as an example, the economy of design, manufacturing and use shall be taken into account at the same time. If the structural scheme of the machine is relatively complex, the design, and manufacturing costs will be relatively higher, but its functions will be more completed, and it will be more productive, so the use economy will be better. On the contrary, for machines with simple structures and incomplete functions, although the design, and manufacturing costs are small, the use costs will be high. When evaluating the design and manufacturing economy of the structural scheme, it can be expressed by the cost of unit efficiency.

It is also necessary to analyze machine's reliability and take it as an evaluation index. Generally speaking, the more complex the system, the lower its reliability of the system. To improve the reliability of complex systems, parallel standby systems must be added, which will inevitably increase the cost of the machine.

Environmental protection is also an important aspect that must be seriously considered in the design. Technical solutions that adversely affect on the environment must be analyzed in detail, and technically mature solutions must be proposed.

The scheme evaluation makes the final decision to determine a schematic diagram or mechanism motion diagram for the following technical design.

At the scheme design stage, we should correctly handle the relationship between reference and innovation. The successful precedents of similar machines should be used for reference, and the original weak links and parts that do not meet the requirements of existing tasks should be improved or fundamentally changed.

3) Technical design

The goal of the technical design phase is to generate general assembly sketches of components. Through sketch design, the shape and basic dimensions of each component and its parts are determined. Based on that, draw the working drawing, assembly drawing, and final assembly drawing of parts.

To determine the dimensions of the part, the following work must be done.

(1) The kinematics design of the machine. According to the determined structural scheme, the parameters of the original parts (power, speed, linear speed, etc.) are determined. Then kinematics calculation is carried out to determine the motion parameters (rotational speed, velocity, acceleration, etc.) of each moving component.

(2) Dynamic calculation of the machine. Combined with the structure and motion parameters of each part, the load and the characteristics of each main part are calculated. The load calculated at this time is only the nominal load acting on the part because the part has not yet been designed.

(3) The design of the working ability of the parts. Given the magnitude and characteristics of the nominal load on the main parts, the preliminary design of parts and components can be made. The working ability criteria are based on the general failure conditions, working characteristics, and environmental conditions of parts and components. Generally, there are strength, stiffness, vibration stability, and service life criteria. The basic dimensions of parts and components can be determined by calculation or analogy.

(4) Design of assembly sketch and general assembly sketch. The assembly sketch and general assembly sketch can be drawn according to the basic dimensions of the main parts and components. In this step, it is necessary to coordinate the structure and size of each part well and consider the structural manufacturability of the designed parts and components comprehensively, to ensure that all parts have the most reasonable configuration.

(5) Check the main parts. The size of all parts is known by the part assembly sketch and general assembly sketch at this stage, and the relationship between adjacent parts is also known. it is difficult to calculate the working capacity in detail for some parts because the specific structure is undetermined in the above steps, so only preliminary calculation and design can be done. We can determine the load acting on the parts and the detailed factors affecting the working ability of the parts. Based on that, it is possible and necessary to accurately check and calculate some important parts with complex shapes and forces. Then repeatedly modify the structure and size of parts until they are satisfactory according to the checking results.

In each step of technical design, optimization design technology can be used to select the best structural parameters. Some new numerical calculation methods, such as the finite element method, can obtain excellent approximate quantitative calculation results for complex problems. The model test method can be used to design some important, complex, and expensive parts. That is, the model is manufactured according to the preliminary design drawings, and the weak parts or redundant cross-section dimensions in the structure are found through tests. Then the original design is modified by strengthening or reducing them until the perfect degree is reached. Reliability evaluation should also be done in the technical design stage, which can evaluate whether the structure and parameters of the designed parts and components meet the reliability requirements, and put forward suggestions for improving the design, thus further improving the design quality of the machine.

After the sketch design is completed, the working drawing of the parts can be designed according to the basic dimensions of the parts determined in the sketch. At this time, there are still a lot of structural details of parts to be scrutinized and determined. When designing working drawings, we should fully consider the machining and assembly manufacturability of parts and the inspection and implementation requirements of parts during and after machining. If some details have an impact on the workability of parts, they must be modified and rechecked. Finally, the working drawings of all parts except standard parts are drawn.

According to the final part working drawing, redraw the assembly drawing, so that we can check out the possible hidden errors in dimensions and structures in the working drawings of parts.

4) Preparation of technical documentation

There are many kinds of technical documents, such as machine design and calculation instructions, operating instructions, standard parts lists, and so on.

When compiling the design calculation specification, all the conclusive contents of scheme selection and technical design should be included.

When compiling the instruction manual of the machine for users, the scope of performance parameters, operation methods, daily maintenance, simple maintenance methods, and the catalog of spare parts should be included.

Other technical documents, such as inspection sheet, list of purchased parts, acceptance conditions, etc., shall be prepared separately as required or not.

5) Application of computer in mechanical design

With the development of computer technology, computers have been widely used in mechanical design, and many high-efficiency designs and analysis softwares have been developed. The software can be used for the comparison of multiple schemes, including large and very complex schemes of structural strength, stiffness, and dynamic characteristics of accurate analysis in the design stage. At the same time, the virtual prototype can be built on the computer, and the design can be verified by virtual prototype simulation to evaluate the feasibility of the design fully. The popularization and application of computer technology in mechanical design have changed the mechanical design process.

The design program of the machine is briefly introduced above. Broadly speaking, in the manufacturing process of a machine, it is possible to modify the design at any time due to technological reasons. If it needs to be modified, certain approval procedures should be followed. After the machine leaves the factory, it should be tracked and investigated in a planned way. Moreover, users will give feedback to manufacturing or design departments. Based on this information, the design department may modify the design after analysis. As a designer, he should have a strong sense of social responsibility, extend his vision to the whole process of manufacturing, using, and even scrapping, and improve the design repeatedly so that the product quality can be continuously improved.

In addition, the above stages are explained respectively.

(1) Planning stage. It is only a preparatory stage. There is only a vague concept of the machine to be designed.

(2) Scheme design stage. This stage determines the success or failure of the design. There are multiple solutions (schemes) in the design work.

The function analysis of a machine is to analyze comprehensively the requirements that must be met, that is, whether the minimum requirements and the expected requirements of the function can be realized, whether there are contradictions among multiple functions, whether they can be replaced by each other, etc. Finally, the functional parameters are determined as the foundation for further design. In this step, we should properly handle the possible contradictions between needs and possibilities, ideals and realities, development goals and current goals, etc.

After the functional parameters are determined, the possible solutions can be put forward. That is, the possible schemes can be proposed. The original, transmission, and executive parts can be considered while seeking the schemes. The more common way is to start the discussion with the executive part.

When discussing the executive part of a machine, the first thing is to choose the working principle. For example, when designing a manufacturing screws machine, the working principle can be either turning threads with a lathe tool on a cylindrical blank or rolling threads with a thread rolling die on a cylindrical blank. Of course, the designed machine will be different according to different working principles. It should be emphasized that new working principles must be constantly researched and developed. This is important to develop design technology.

2.2.3 Mechanical Design Steps

Before the design, it is necessary to formulate the design tasks. When designing a complex task, there are generally three design stages: preliminary design, technical design, and working diagram design. The preliminary design includes determining the working principle and basic structural form, motion design, designing the main parts and components, and drawing and reviewing the preliminary design. The technical design includes modifying the design according to the review opinions, designing all parts and components, and drawing and reviewing the technical design. The working diagram design includes revising the design according to the review opinions, drawing all working diagrams, and formulating all technical documents. In each step of design, it is possible to find some unreasonable decisions in the previous step. It is necessary to go back to the previous step, modify the unreasonable design and then redo the subsequent work.

When the design task is relatively simple, such as simple machinery, inheritance design, or variant design of general machinery, the design can directly start from the technical design, and the work drawing design will be carried out after examination, modification, and approval, thus becoming a two-stage design. For batch or mass-production products, it is necessary to carry out stereotype design.

The following steps are needed to follow.

(1) Make design tasks.

The design task is based on user orders, market needs, and new scientific research achievements. The design department shall draw up possible plans using various technologies and market intelligence, compare their advantages and disadvantages, consult with the business department and users, and make reasonable design tasks and objectives, which is particularly important for new designs. The failure of mission objectives will result in serious economic loss and even total failure.

(2) Determine the working principle and basic structure.

If the design task is not specified, the first step of the design is to determine the overall scheme, that is, the working principle and the corresponding structural form. For example, when designing a high-power marine diesel engine, a two-stroke, double-acting, crosshead, low-speed diesel engine or a four-stroke, single-acting, medium-speed diesel engine has to be selected. When designing a crushing machine for coarse rock crushing, the jaw or rotary crusher with extrusion and bending as the main crushing action, or the single rotor or double rotor impact crusher with impact as the main crushing action has to be selected.

(3) Motion design.

After the design scheme is determined, it is necessary to select the appropriate mechanism to obtain the required motion. The jaw crusher mentioned above breaks the rock entering the crushing chamber by squeezing, bending, and splitting action depending on the swing of its moving jaw plate, while the swing of the moving jaw plate can be a simple swing of a double toggle mechanism or a complex swing of a single toggle mechanism. In a new design, synthesizing a new mechanism to obtain the required motion scheme is often a difficult task. Therefore, designers can use the motion scheme provided by existing and mature mechanisms for reference.

(4) Structural design.

Structural design includes calculating the force, strength, size, and weight of the main parts of the machine and drawing sketches of the main parts and components. The designer will find contradictions in the shape, size, proportion, and other aspects of each part by sketching. At the same time, the designer should check whether there are parts that may lead to overheating, excessive wear, or vibration. The original structure must be adjusted or modified, if it is not feasible. It should be mentioned that strengthening or improving one aspect may weaken or deteriorate the other aspect. We must balance and coordinate to achieve the best comprehensive effect. The sketch will be modified repeatedly until preliminarily satisfactory. Then the preliminary general drawing and cost estimation can be made. The preliminary general drawing shall be drawn strictly to scale, and sufficient views and sections shall be selected.

(5) Draw the working diagram.

After the final modification is made according to the opinions of the second review, formal part drawings, component assembly drawings, and general assembly drawings can be drawn. There are two tasks during part drawing. One is the process audit which can ensure the parts are easy to produce and reduce the manufacturing cost. The other is the standard review to make sure that the structural elements, dimensions, tolerance fit, heat treatment technical conditions, standards, and general parts conform to the provisions of the standard.

It is very important to check the drawings after the completion of the part drawings. The designer shall coordinate the dimensions between parts and check the tolerance and fit between coupling parts. The carefully checked drawings can ensure smooth assembly after processing. The most reliable proofreading method is to redraw a general assembly drawing based on the part drawing, and all contradictions will be displayed.

(6) Trial production and finalized design.

The design drawings completed by the above steps can be put into formal production for single-piece or small-batch production machinery. For batch or mass production machinery, the prototype shall be trial produced before formal production, and the functional test and identification shall be performed. After passing the test, the batch trial production shall be carried out according to the batch production process. Problems in batch trial production may also require corresponding design modifications before it becomes a finalized design for formal production.

2.3 Mechanical Design Criteria

2.3.1 Technical Performance Criteria

Technical performance includes all performances of product functions, manufacturing, and operation conditions. It refers to both static performance and dynamic performance. For example, the power, efficiency, service life, strength, stiffness, anti-friction, wear performance, vibration stability, thermal characteristics, etc., which the product can transmit. Technical performance criteria refer to that the relevant technical performance must meet the specified requirements. For example, vibration will generate additional dynamic load and variable stress, especially when its frequency is close to the natural frequency of the mechanical system or parts. Resonance will occur, and then the amplitude will increase sharply, which may lead to rapid damage on components or even the whole system. The vibration stability criterion is to limit the relevant vibration parameters of a mechanical system or parts, such as natural frequency, amplitude, noise, etc., within the specified allowable range. For another example, the heating of the machine during operation may cause thermal stress, thermal strain, and even thermal damage. The thermal characteristic criterion is to limit various relevant thermal parameters, such as thermal stress, thermal strain, temperature rise, etc., within the specified range.

2.3.2 Standardization Criteria

The primary standards related to mechanical product design are as follows.

(1) Concept standardization. Terms, symbols, measurement units, etc., involved in the design process shall conform to the standards.

(2) Standardization of physical form. The structural form, size, performance of parts, raw materials, equipment, and energy shall be selected according to uniform provisions.

(3) Method standardization. Operation methods, measurement methods, test methods, etc., shall be implemented according to the corresponding provisions.

The standardization criterion is that all behaviors in the whole design process should meet the above standardization requirements. The published standards related to mechanical part design can be divided into national standards, industrial standards, and enterprise standards in terms of application scope. From respective of mandatory use, it can be divided into mandatory use and recommended use.

2.3.3 Reliability Criteria

Reliability is defined as the probability that a product or component can complete a specified function within its expected life under specified service conditions. Reliability criteria mean that the designed products, components, or parts shall meet the specified reliability requirements.

2.3.4 Safety Criteria

The safety of the machine includes the following aspects.

(1) Part safety. It means that the parts will not occur fracture, excessive deformation, excessive wear, or loss of stability under the specified external load and within the specified time.

(2) Safety of the whole machine. It refers to the requirement that the machine will not fail under specified conditions and can normally realize the overall function.

(3) Work safety. It refers to the protection of operators to ensure personal safety, and physical and mental health.

(4) Environmental safety. It means that the environment around the machine and people will not be polluted or harmed.

Chapter 3 Typical Mechanical Parts

3.1 Gears

Gears are toothed, cylindrical wheels that transmit motion and power from one rotating shaft to another. The teeth of a driving gear mesh accurately in the spaces between teeth on the driven gear. The driving teeth push the driven teeth, applying a force perpendicular to the radius of the gear. Therefore, a torque is transmitted, and power is transmitted by the rotating gear.

3.1.1 Involute Spur Gear

The most widely used spur gear tooth form is the full-depth involute form. Its characteristic shape is shown in Figure 3-1.

The involute is one of a class of geometric curves called conjugate curves. When two involute spur gear teeth are in mesh and rotating, they have a constant angular velocity ratio. From the moment of initial contact to the moment of disengagement, the speed of the driving gear is in constant proportion to the speed of the driven gear. The resulting action of the two gears is very smooth. Otherwise, there would be speeding up or slowing down during the engagement, with the resulting accelerations causing vibration, noise, and dangerous torsional oscillations in the system.

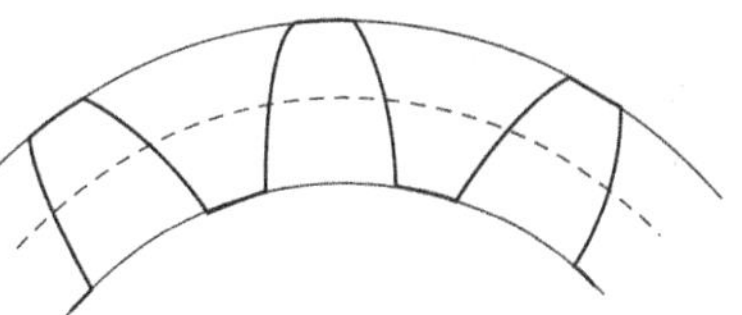

Figure 3-1 Involute-tooth form

An involute curve can be seen by wrapping a string around a cylinder's circumference. Tie a pencil to the end of the string, then start with the pencil tight against the cylinder and hold the string taut. Move the pencil away from the cylinder while keeping the string taut. The curve drawn is an involute. The circle represented by the cylinder is called the base circle. Notice that the string is tangent to the base circle and perpendicular to the involute at any position on the curve. Drawing another base circle along the same centerline at the point where the resulting involute is tangent to the first one. It demonstrates that the line tangent to the base circles is identical to the meshing line when two gear teeth are in mesh.

It is a fundamental principle of kinematics and the study of motion that if the line drawing perpendicular to the surfaces of two rotating bodies at the contact point always crosses the centerline of the two bodies at the same place, and the angular velocity ratio of the two bodies will be constant. As demonstrated here, the gear teeth of involute spur gear obey the law.

3.1.2 Involute Helical Gear

Helical and spur gears are distinguished by the orientation of their teeth. For the spur gear, the teeth are straight and aligned with the axis of the gear. For the helical gears, the teeth are inclined at an angle with the axis, this angle is called the helix angle. Figure 3-2 shows two examples of helical gear.

Figure 3-2 Helical gears

The forms of helical gear teeth are very similar to those discussed for spur gears. The basic parameter is the helix angle. Typical helix angles range from approximately 10° to 30°, but angles up to 45° are practical.

The helix for a given gear can be either left-hand or right-hand. In a typical installation, helical gears would be mounted on parallel shafts. It is required that one gear should be of the right hand and the other be left hand with an equal helix angle. If both meshing gears are of the same hand, the shafts will be 90 degrees from each other. Such gears are called crossed helical gears. Parallel shaft arrangement for helical gears is preferred because it makes a much higher power-transmitting capacity than the crossed helical arrangement for a given size.

The main advantage of helical gears over spur gears is the smoother engagement, because a given tooth bears its load gradually, not suddenly. Contact starts at one end of a tooth near the tip and progresses across the face in a path down across the pitch line to the lower flank of the tooth where it leaves engagement. Simultaneously, other teeth are coming into engagement before a given tooth leaves engagement, increasing the average number of teeth engaged and sharing loads compared with a spur gear. The lower average load per tooth allows a greater power transmission capacity for a given gear size , or a smaller gear can be designed to carry the same power.

The main disadvantage of helical gears is that an axial thrust load is produced as a natural result of the inclined arrangement of the teeth. The bearings that hold the shaft carrying the helical gear must be capable of reacting against the thrust load.

3.1.3 Bevel Gears

Bevel gears have teeth arranged as elements on the surface of a cone. The teeth of straight bevel gears appear to be similar to spur gear teeth, but they are tapered, being wider at the outside and narrower at the top of the cone. Bevel gears typically operate on shafts that are 90° from each other. Indeed, this is the reason bevel gears are used in a drive system. Especially designed bevel gears can operate on shafts whose angle is not 90°. When bevel gears are made with teeth in a helix angle similar to that in helical gears, they are called spiral bevel gears. They operate more smoothly than spur bevel gears and can be sized smaller for a given power transmission capacity. When both bevel gears in a pair have the same number of teeth, they are called miter gears and are only used to change the axes of the shaft to 90 degrees. No speed change occurs. Figure 3-3 shows an example of bevel gear.

3.1.4 Rack and Pinion Drive

A rack is a straight gear that moves linearly instead of rotating. When a circular gear is mated with a rack, the combination is called a rack and pinion drive. It is always applied to the steering mechanism of a car or a part of other machinery. Figure 3-4 shows an example of a rack and pinion drive.

Figure 3-3 Bevel gear

Figure 3-4 Rack and pinion drive

3.1.5 Worm Drive

A worm and its mating worm gear operate on shafts that are 90 degrees from each other. They typically accomplish a rather large speed reduction ratio compared with other types of gear. The worm is the driver, and the gear is the driven part. The teeth on the worm appear similar to screw threads; indeed, they are often called threads rather than teeth. The teeth of the worm gear can be straight like spur gear teeth or can be helical. One disadvantage of the worm drive is that it has lower mechanical efficiency than most other kinds of gears on account of the extensive rubbing contact between the surfaces of the worm threads and the sides of the worm gear teeth. Figure 3-5 shows an example of a worm drive.

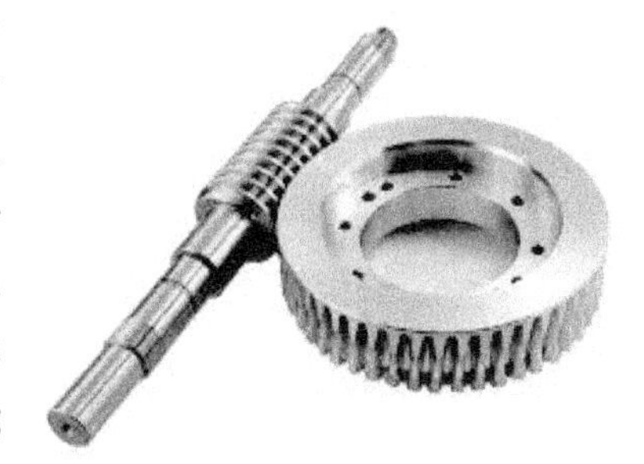

Figure 3-5 Worm drive

3.2 Shaft

A shaft is usually a relatively long part that rotates and transmits power. One or more components, such as gears, sprockets, pulleys, and cams, are usually connected to the shaft by pins, keys, splines, snap rings or other devices. These components mentioned above are "related parts," as well as couplings and clutches, which are used to realize the connection or separation between the shaft and power source or load.

The shaft is one of the main supporting parts of the machine. The transmission parts (such as gears, worm gears, pulleys, sprockets, rollers, etc.) that do rotary motion must be installed on the shaft to transmit motion and power. The shaft will bear various combined forces of axial, bending, torsion, and other loads, and these loads may be static or fluctuating.

3.2.1 Types of Shaft

1) The classification of the shaft according to the load

(1) Spindle.

It is used to install and support roller, rope pulley, wheel, and other parts and bear the bending moment generated by radial load without torque. The spindle can be divided into the fixed spindle and the rotating spindle. The parts supported on the fixed mandrel can rotate freely. The mandrel is the only subject to static or pulsating bending stress, as shown in Figure 3-6a), while the rotating mandrel and its upper parts are fixed and rotate together and subject to alternating bending stress, as shown in Figure 3-6b). From the perspective of stress, the fixed mandrel should be preferred because it has better stress conditions. From the support point of view, it is easier to assemble, disassemble, clean, and lubricate the rotating mandrel than the fixed mandrel.

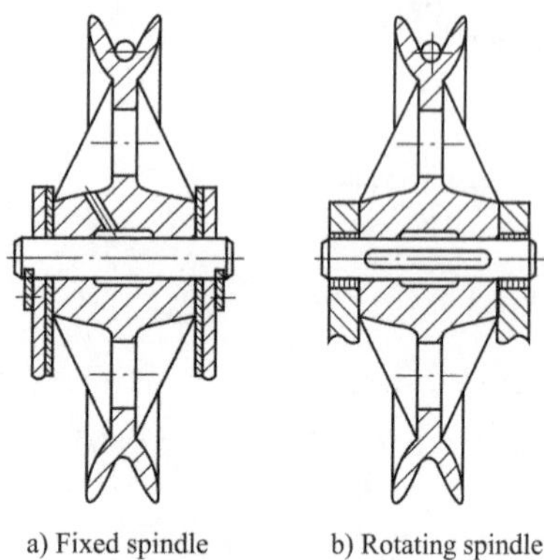

a) Fixed spindle b) Rotating spindle

Figure 3-6 Spindle

(2) Rotating shaft.

Generally, they rotate and transmit torque all the way around. Torque inputs and outputs through gears, pulleys, couplings, and other parts. As shown in Figure 3-2, the power inputs from the shaft end are transmitted to the gear through the shaft. During operation the shaft bears the effect of bending moment and torque simultaneously.

(3) Transmission shaft.

The shaft mainly bears torque with little or no bending moment. As shown in Figure 3-3, the transmission shaft of the vehicle is used to transmit torque and motion from the gearbox to the rear axle. This shaft mainly bears torque.

2) The classification of the shaft according to shape

(1) Straight shaft.

The axis of each shaft section is in the same line, as shown in Figure 3-6 to Figure 3-8. The straight shaft can be divided into the smooth shaft and the stepped shaft. The shape of the smooth shaft is simple, but the parts on the shaft are not easy to locate. It is often used for the spindle and transmission shaft, as shown in Figure 3-6 and Figure 3-8. The stepped shaft is easy to locate, so it is often used for the rotating shafts, as shown in Figure 3-7.

Figure 3-7 Rotating shaft

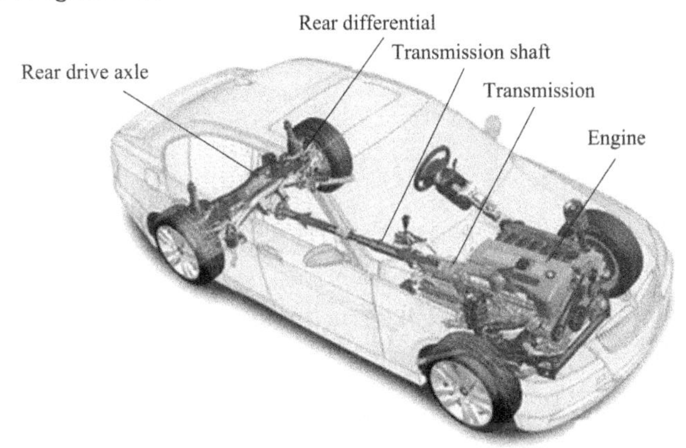

Figure 3-8 Transmission shaft

(2) Crankshaft.

The axes of each shaft segment of the crankshaft are not in the same straight line. The crankshaft is often used in reciprocating motion mechanisms, such as the crankshaft in internal combustion engines, as shown in Figure 3-9.

Figure 3-9 Crankshaft

(3) Flexible shaft.

The flexibal shaft is always used for small power machinery with main driving position change, such as manual grinder, or some equipments with fixed position but serious dislocation with the prime mover (such as tachometer). The common flexible shaft is made of multiple groups of steel wires wound in layers, with movable metal protective sleeves outside, as shown in Figure 3-10a). It has good flexibility and can transmit torque and rotating motion to the required position flexibly, as shown in Figure 3-10b). The torque transmitted by the flexible shaft is related to the diameter of the shaft core, the type of structure, the length, the minimum bending radius, and other factors.

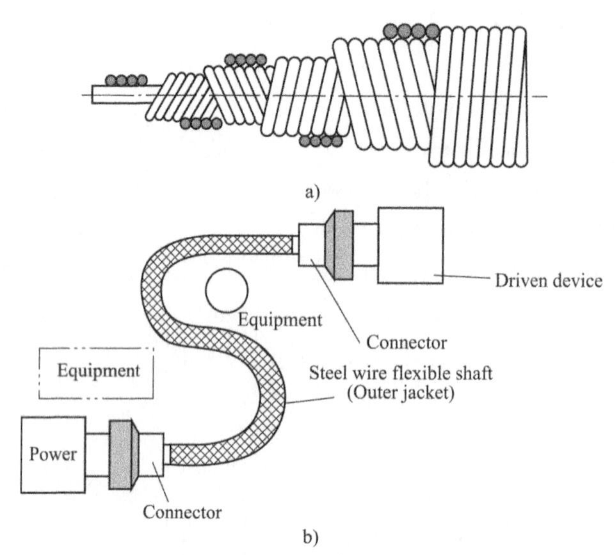

Figure 3-10 Steel wire flexible shaft

3.2.2 Material and Heat Treatment of Shaft

The material of the shaft should first have sufficient strength, low sensitivity to stress concentration, and then meet the requirements of stiffness, wear resistance, corrosion resistance, and other aspects, and have good processing technology, low price, and be easy to obtain. The shaft material shall be selected according to the specific conditions by the principles of economy, rationality, and applicability.

Common materials of the shafts are mainly carbon steel and alloy steel. The billet of steel shaft is mostly rolled round steel and forgings, followed by nodular cast iron and high-strength cast iron.

Carbon steel is cheaper than alloy steel, the sensitivity to stress concentration is relatively lower, and its strength and wear resistance can be improved by heat treatment and chemical heat treatment. It is convenient for mechanical processing, so carbon steel is particularly widely used for manufacturing. 45 steel is the most commonly used in general machines. For shafts with small or unimportant stress, they can also be made of Q235, Q255, Q275, and other ordinary carbon steel.

Alloy steel has higher mechanical properties and superior quenching properties than carbon steel. However, the stress concentration is relatively sensitive, and the price is more expensive. It is often used for shafts with high manufacturing strength and wear resistance requirements or other special requirements, such as high-speed and heavy-duty shafts, or shafts with large stress and requirements to minimize the size and weight, as well as those working under high temperature, low temperature or corrosion conditions. Since the difference in elastic modulus between alloy steel and carbon steel is very small at normal working temperature (below 20℃), replacing carbon steel with alloy steel cannot improve the shaft rigidity remarkably.

Various heat treatments, such as high-frequency quenching, carburizing, nitriding, cyanidation, etc., and surface strengthening treatments, such as shot peening and rolling significantly improve the fatigue strength of the shaft.

Nodular cast iron and high-strength cast iron have superior technological properties, which are suitable for manufacturing shafts with complex shapes (such as camshaft, crankshaft, etc.), and have the advantages of low price, good vibration absorption, wear resistance, and low sensitivity to stress concentration, but the quality of cast shafts is not easy to control, and the reliability is poor.

3.2.3 Shaft Design

1) Shaft design requirements

Generally, the working capacity of the shaft depends on its strength and rigidity. It also depends on its vibration stability for the high-speed shaft. Due to the different working requirements of various machines for shafts, the main design problems are different. For the general shaft, it is required to have sufficient strength, rigidity, reasonable structure, and good processability. Therefore, strength calculation should be carried out to prevent fracture. Rigidity calculation shall also be carried out for shafts with rigidity requirements, such as lathe spindle, to prevent excessive deformation during operation. In addition, the vibration stability calculation should be carried out to prevent resonance damage. The shaft's structural design shall realize the shaft's bearing function and ensure that it is supported, ensure the mutual position of all parts, and transmit the torque smoothly. At the same time, it shall have a reasonable structure and good technology.

2) Design steps of shaft

There is no fixed step for shaft design. It should be determined according to specific conditions. It rarely calculates the diameter section by section in the actual design process of the spindle and the rotating shaft. Instead, the minimum diameter obtained from the required diameter is first determined by the structural design method. Then, the static strength, fatigue strength, and allowable deformation (stiffness) should be checked. If necessary, the critical speed should be checked equally. The general design steps are:

(1) Determine the position of parts on the shaft according to the transmission scheme.

(2) Select the material of the shaft.

(3) Preliminarily estimate the diameter of the shaft.

(4) Carry out the structural design of the shaft.

(5) Check the strength of the shaft.

(6) Check and calculate the rigidity and critical speed of the shaft if necessary.

(7) Draw the working diagram of the shaft.

The above steps are often crossed and repeated. When the shaft's strength, rigidity, or stability does not meet the requirements, the material, or structure of the shaft shall be analyzed and calculated again. In addition, some parts related to the shaft, such as the box, bearings, couplings, keys, etc., often need to be designed simultaneously and coordinated with the shaft design.

3) Preliminary estimation of shaft diameter

When designing a shaft, the minimum diameter of the shaft is usually estimated first as the basis for structural design. It is usually estimated according to the strength condition. The calculation formula is:

$$d \geqslant \sqrt[3]{\frac{9550 \times 10^3}{0.2[\tau_T]}} \sqrt[3]{\frac{P}{n}} \geqslant A \sqrt[3]{\frac{P}{n}} \text{mm} \tag{3-1}$$

Where, d——the diameter of the shaft at the specified section, mm;

n——the shaft speed, r/min;

P——the power transmitted by the shaft, kW;

$[\tau_T]$——the allowable torsional shear stress, MPa;

A——the computing constant related to the shaft material, which can be determined by:

$$A = \sqrt[3]{\frac{9550 \times 10^3}{0.2[\tau_T]}} \tag{3-2}$$

4) Structural design of shaft

The shaft comprises the journal, shaft head, and shaft body. The shaft-supported part is called the journal, the hub-installed part is called the shaft head, and the part connecting the journal and the shaft head is called the shaft body. The diameter of the journal and shaft head shall be the standard value of the standard circle, especially the journal with rolling bearing, which must be selected according to the bore diameter of the bearing.

When considering the structure of the shaft, the following requirements shall be met:

(1) The stress of the shaft is reasonable to improve its strength and rigidity of the shaft.

(2) The parts installed on the shaft shall be fixed firmly and reliably (axial and circumferential fixation).

(3) The structure on the shaft shall be easy to process, assemble, disassemble, adjust, and minimize stress concentration.

5) Strength check calculation of shaft

Once the size of the shaft is preliminarily determined through the structural design, the strength of the shaft can be checked and calculated according to the load condition. For general steel shafts, the equivalent stress σ_e of the dangerous section can be calculated by the third strength theory. The formula is:

$$\sigma_e = \sqrt{\sigma_b^2 + 4\tau_T^2} \leqslant [\sigma_b] \tag{3-3}$$

Where, σ_e——the bending stress, MPa;

τ_T——the torsional shear stress, MPa;

$[\sigma_b]$——the allowable bending stress, MPa.

3.3 Rolling Contact Bearings

The purpose of a bearing is to support a load while permitting relative motion between two elements of a machine. Rolling contact bearings refer to the wide variety of bearings that use spherical balls or some other type of roller between the stationary and the moving elements. The most common type of bearing supports a rotating shaft, resisting purely radial loads or a combination of radial and axial (thrust) loads. Some bearings are designed to carry only thrust loads.

The components of a typical rolling contact bearing are the inner race, the outer race, and the rolling elements. Figure 3-11, Figure 3-12 show the common single-row, deep-groove ball bearing. Usually, the outer race is stationary and is held by the machine's housing. The inner race is pressed onto the rotating shaft and thus rotates with it. Then, the balls roll between the outer and inner races. The load path is from the shaft to the inner race, then to the balls, next to the outer race, and finally to the housing. The balls' presence allows a very smooth, low-friction rotation of the shaft. The typical coefficient of friction for a rolling contact bearing is approximately 0.001 to 0.005.

Figure 3-11 The common single-row bearing

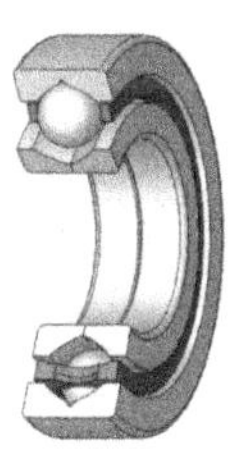

Figure 3-12 The common deep-groove ball bearing

3.3.1 Types of Rolling Contact Bearing

Here, we will discuss several different types of rolling contact bearings and their typical applications, they are the single-row deep-groove ball bearings, double-row deep-groove ball bearings, and angular contact ball bearings.

Radial loads act toward the center of the bearing along a radius. Radial loads are typical of those created by power transmission elements on shafts such as spur gears, V-belt drives, and chain drives. Thrust loads are those that act parallel to the axis of the shaft. The axial components of the forces on helical gears, worms, worm gears, and bevel gears are thrust loads. Also, bearings supporting shafts with vertical axes are subjected to thrust loads due to the weight of the shaft and the elements on the shaft, as well as from axial operating forces. Misalignment refers to the angular deviation of the axis of the shaft at the bearing from the actual axis of the bearing itself. An excellent rating for misalignment in Table 3-1 indicates that the bearing can accommodate up to 4.0° of angular deviation. A bearing with a fair rating can withstand up to 0.15°, while a poor rating indicates that rigid shafts with less than 0.05° of misalignment are required. Manufacturers' catalogs should be consulted for specific data.

An excellent rating for the misalignment of different bearings Table 3-1

Bearing types	Radial load capacity	Thrust load	Misalignment capability
Single-row, deep-groove ball	Good	Fair	Fair
Double-row, deep-groove ball	Excellent	Good	Fair
Angular contact	Good	Excellent	Poor
Cylindrical roller	Excellent	Poor	Fair
Needle	Excellent	Poor	Poor
Spherical roller	Excellent	Fair/good	Excellent
Tapered roller	Excellent	Excellent	Poor

(1) Single-Row, Deep-Groove Ball Bearing.

The inner race is typically pressed on the shaft at the bearing seat with a slight interference fit to ensure that it rotates with the shaft. The spherical rolling elements, or balls, roll in a deep groove in both the inner and the outer races. The spacing of the balls is maintained by retainers or "cages". The single-row, deep-groove ball bearing is shown in Figure 3-11. While designed primarily for radial load-carrying capacity, the deep groove allows a fairly sizable thrust load to be carried. The thrust load would be applied to one side of the inner race by a shoulder on the shaft. The load would pass across the side of the groove, through the ball, to the opposite side of the outer race, and then to the housing. The radius of the ball is slightly smaller than the radius of the groove to allow the free rolling of the balls. The contact between a ball and the race is theoretically at a point, but it is a small circular area because of the deformation of the elements. Because the load is carried on a small area, very high local contact stresses will occur. A bearing with a greater number of balls or larger balls operating in larger-diameter races should be used to increase the capacity of a single-row bearing.

(2) Double-Row, Deep-Groove Ball Bearing.

Adding a second row of balls (Figure 3-13) increases the radial load-carrying capacity of the deep-groove type of bearing compared with the single-row design because more balls share the load. Thus, a greater load can be carried in the same space, or a given load can be carried in a smaller space. The greater width of double-row bearings often adversely affects the misalignment capability.

(3) Angular Contact Ball Bearing.

One side of each race in an angular contact bearing is higher to accommodate greater thrust loads than the standard single-row and deep-groove bearing. Figure 3-14 shows the preferred angle of the resultant force (radial and thrust loads combined), with commercially available bearings having angles of 15° to 40°.

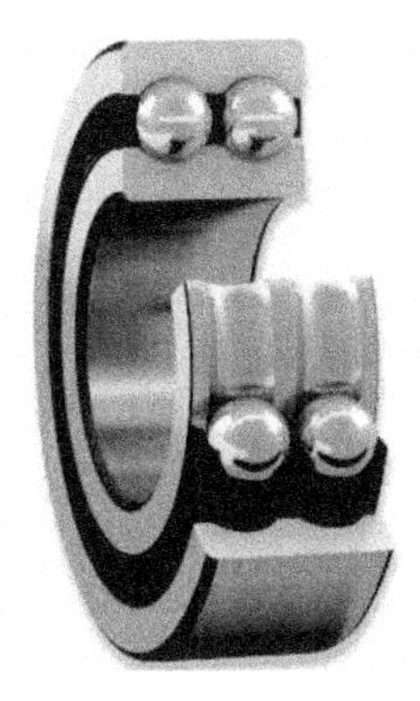

Figure 3-13 Double-row deep groove ball bearing

Figure 3-14 Angular contact ball bearing

3.3.2 Bearing Materials

The load on a rolling contact bearing is exerted on a small area. The resulting contact stresses are pretty high regardless of the type of bearing. Contact stresses around 300000psi are not uncommon in commercially available bearings. The balls, rollers, and races are made from tough, high-strength steel or ceramic to withstand such high stresses.

Rolling elements and other components can be made of ceramic materials, such as silicon nitride. Although the cost is higher than that of steel, ceramics offer several significant advantages like light weight, high strength, and high-temperature capability, making them desirable for aerospace, engine, military, and other demanding applications.

1) Practical considerations in the application of bearings

The functions of lubrication in a bearing unit are as follows.

(1) To provide a low-friction film between the rolling elements and the races of the bearing and at points of contact with cages, guiding surfaces, retainers, and so on.

(2) To protect the bearing components from corrosion.

(3) To help dissipate heat from the bearing unit.

(4) To carry heat away from the bearing unit.

(5) To help dispel contaminants and moisture from the bearing.

Rolling contact bearings are usually lubricated with grease or oil. Grease is satisfactory for the normal ambient temperatures (approximately 70 ℉) and relatively slow speeds (under 500rpm) condition. Continuous oil lubrication is required for higher speeds or higher ambient temperatures, possibly with external cooling for the oil.

Oils used in bearing lubrication are usually clean, stable mineral oils. For lighter loads and lower speeds, light oil is used. Heavier loads and higher speeds require heavier oils up to SAE30. A recommended upper limit for lubricant temperature is 160T. It depends on the operating temperature of the lubricant in the bearing. Manufacturers' recommendations should also be considered.

In some critical applications, such as bearings in jet engines and very-high-speed devices, lubricating oil is pumped under pressure to an enclosed housing for the bearing, where the oil is directed at the rolling elements themselves. The temperature of the oil in the sump is monitored and controlled with heat exchangers or refrigeration to maintain the oil viscosity within acceptable limits. Such systems provide reliable lubrication and ensure the removal of heat from the bearing.

Greases used in bearings are mixtures of lubricating oils and thickening agents. The fatty acid soap made of lithium or barium are most commonly used thickening agent. It acts as carriers for the oil, which is drawn out at the point of need within the bearing. Additives to resist corrosion or oxidation of the oil sometimes need to be added. Classifications of greases specify the operating temperatures to which the greases will be exposed, as defined by the American Bearing Manufacturers Association (ABMA).

2) Installation

The interference fit between the bearing hole and the shaft precludes the possibility of rotation of the inner race of the bearing for the shaft. Such a condition would result in uneven wear of the bearing elements and early failure. Installing the bearing then requires rather heavy forces applied axially. Care must be exercised so that the bearing is not damaged during installation. The installation force should be applied directly to the inner race of the bearing through the rolling elements to the inner race. Because of the small contact area, such transmission of forces would likely overstress some elements and exceed the static load capacity. This will lead to brinelling, along with the noise and rapid wear that accompany this condition. For large bearings, it is necessary to heat the bearing to expand its diameter to keep the installation forces within a reasonable situation. Removal of bearings intended for reuse requires similar precautions. Bearing pullers are available to facilitate this task.

3) Preloading

Some bearings are made with internal clearances that must be taken up in a particular direction to ensure satisfactory operation. In such cases, preloading must be provided, usually in the axial direction. On horizontal shafts, springs are typically used, with axial adjustment of the spring deflection sometimes provided to adjust the amount of preload. When space is limited, the use of Belleville washers is desirable because they provide high forces with small deflections. Shims can be used to adjust the actual deflection and preload obtained. On vertical shafts, the weight of the shaft assembly itself may be sufficient to provide the required preload.

4) Sealing

When the bearing operates in dirty or moist environments, special shields and seals are usually specified. They can be provided on either or both sides of the rolling elements. Shields are typically metal and are fixed to the stationary race. Seals are made of elastomeric materials and contact with the rotating race. The bearings fitted with seals and shields and precharged at the factory with grease are sometimes called permanently lubricated. Although such bearings are likely to give many years of satisfactory service, extreme conditions can produce degradation of the lubricating properties of the grease. The presence of seals also increases the friction in a bearing. Sealing can be provided outside the bearing in the housing or at the shaft/housing interface. On high-speed shafts, a labyrinth seal, consisting of a non-contacting ring around the shaft with a few thousandths of an inch radial clearance, is frequently used. Grooves, sometimes in the form of a thread, are machined in the ring. The relative motion of the shaft with respect to the ring creates the sealing action.

5) Limiting speeds

Most mechanical engineering catalogs will list the limitation of speed for each bearing. It may result in excessively high operating temperatures due to friction between the cages supporting the rolling elements if exceeding the limitations. Generally, the limiting speed is lower for larger bearings than for smaller bearings. Also, a given bearing will have a lower limiting speed as loads increase. With special care, either in the fabrication of the bearing cage or in the lubrication of the bearing, bearings can be operated at higher speeds than those listed in the catalogs. In such applications, it needs to consult the manufacturer. Ceramic rolling elements with lower mass can work at higher limiting speeds.

3.4 Keys, Couplings, and Seals

A key is a machinery component that provides a torque transmitting link between two power-transmitting elements. Coupling refers to a device used to connect two shafts together at their ends for transmitting power. A mechanical seal is simply a method of containing fluid within a vessel (typically pumps, mixers, etc.) where a rotating shaft passes through a stationary housing or occasionally, where the housing rotates around the shaft.

3.4.1 Square and Rectangular Parallel Keys

The key is demountable to facilitate the assembly and disassembly of the shaft system (Figure 3-15b). It is installed in an axial groove machined into the shaft, called a keyseat. A similar groove in the hub of the power-transmitting element is usually called a keyseat. The key is typically installed into the shaft keyseat first; then, the hub keyseat is aligned with the key, and the hub is slid into position. Parallel keys are most widely utilized. They have a square or rectangular cross-section. The most common type of key for shafts up to 6.5 inches in diameter is the square key, as illustrated in Figure 3-15a). The rectangular key illustrated in Figure 3-15b) is recommended for larger shafts and is used for smaller shafts where the shorter height can be tolerated. Both the square and the rectangular keys are referred to as parallel keys because the top, bottom, and sides of the key are parallel.

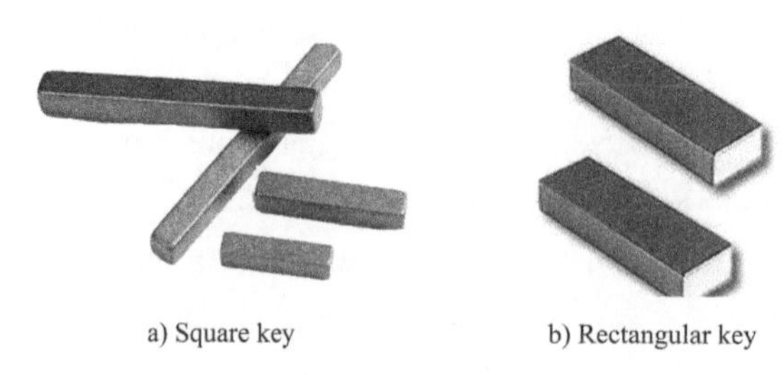

a) Square key　　　b) Rectangular key

Figure 3-15　Square and rectangular parallel keys

The keyseats in the shaft and the hub are designed so that exactly one-half of the height of the key is bearing on the shaft keyseat side and the other half on the hub keyseat side.

3.4.2 Couplings

The term coupling refers to a device used to connect two shafts at their ends to transmit power. There are two general types of couplings, respectively rigid and flexible.

(1) Rigid Couplings.

Rigid couplings are designed to draw two shafts together tightly so that no relative motion can occur between them. This design is desirable for certain kinds of equipment where precise alignment of two shafts is required and can be provided. In such cases, the coupling must be designed to be capable of transmitting the torque in the shafts.

Figure 3-16　Rigid coupling

A typical rigid coupling is shown in Figure 3-16, in which flanges are mounted on the ends of each shaft and are drawn together by a series of bolts. The load path is from the driving shaft to its flange, through the bolts, into the mating flange, and out to the driven shaft. The torque places the bolts in shear. The total shear force on the bolts depends on the radius of the bolt circle and the torque.

Rigid couplings should be only used when the alignment of the two shafts can be maintained very accurately for both the installation and operation of the machines. If significant angular, radial, or axial misalignment occurs, stresses that are difficult to predict may lead to early failure in the shafts. These difficulties can be overcome by the use of flexible couplings.

(2) Flexible Couplings.

Flexible couplings are designed to transmit torque smoothly while permitting some axial, radial, and angular misalignment. The flexibility is that when misalignment does occur, parts of the coupling move with little or no resistance. Thus, no significant axial or bending stresses are developed in the shaft.

Many types of flexible couplings are available commercially. Each of them is designed to transmit a given limiting torque. The manufacturer's catalog lists the design data from which one can choose a suitable coupling. Remember that torque equals power divided by rotational speed. So, for a given size of coupling, as the speed of rotation increases, the amount of power the coupling can transmit also increases, although not always in direct proportion. Of course, centrifugal effects determine the upper limit of speed.

3.4.3 Seals

Seals are an important part of machine design in situations in the following conditions.

(1) Contaminants must be excluded from critical areas of a machine.

(2) Lubricants must be contained within a space.

(3) Pressurized fluids must be contained within a component such as a valve or a hydraulic cylinder.

(4) Some of the parameters affecting the choice of the type of sealing system, the materials used, and the details of its design are as follows.

① The nature of the fluids to be contained or excluded.

② Pressures on both sides of the seal.

③ The nature of any relative motion between the seal and the mating components.

④ Temperatures on all parts of the sealing system.

⑤ The degree of sealing required. Is some small amount of leakage permissible?

⑥ The life expectancy of the system.

⑦ The nature of the solid materials against which the seal must act, such as corrosion potential, smoothness, hardness, and wear resistance.

⑧ Ease of service for replacement of worn sealing elements.

The number of designs for sealing systems is virtually limitless, and only a general idea is presented here. Often, designers rely on technical information provided by manufacturers of complete sealing systems or specific sealing elements. Also, testing of a proposed design is advised in critical or unusual situations.

3.5 Belt and Chain Drives

Belts and chains represent the major types of flexible power transmission elements. The belt drive is a frictional type of mechanical drive where friction force between belt and pulley is used to transmit power and motion. Chain drive is one engagement type mechanical drive where power and motion are transmitted by successive engagement and disengagement of chain with a sprocket.

3.5.1 Belt Drives

1) Structural characteristics of belt drives

The electric motor develops rotary power, but motors typically operate at very high speeds and deliver very low torque to be appropriate for the final drive application. Remember, for a given power transmission, the torque is increased in proportion to the amount that rotational speed is reduced. The high speed of the motor makes belt drives somewhat ideal for that first stage of reduction. A smaller drive pulley is attached to the motor shaft, while a larger diameter pulley is attached to a parallel shaft that operates at a correspondingly lower speed. Pulleys for belt drives are also called sheaves.

When the belt is used for speed reduction, the smaller sheave is mounted on the high-speed shaft, such as the shaft of an electric motor. The larger sheave is mounted on the driven machine.

The belt is installed around the two sheaves while the center distance between them is reduced. Then, the sheaves are moved apart, placing the belt in a rather high initial tension. When the belt is transmitting power, the friction causes the belt to grip the driving sheave, increasing the tension on one side, called the "tight side". The tensile force in the belt exerts a tangential force on the driven sheave; thus, a torque is applied to the driven shaft. The opposite side of the belt is still under tension but at a smaller value, which is called the "slack side".

The linear speed of a belt is usually 2500 ~ 6500ft/min, which results in relatively low tensile forces in the belt. At lower speeds, the tension in the belt becomes too large for typical belt cross sections, and slipping may occur between the sides of the belt and the sheave or pulley. At higher speeds, dynamic effects such as centrifugal forces, belt whip, and vibration reduce the effectiveness of the drive and its life. The speed of 4000ft/min is generally recommended. Some belt designs employ high-strength, reinforcing strands, and some cogged designs engage matching grooves in the pulleys to enhance their ability to transmit high forces at low speeds.

2) Types of belts

Many types of belts are available, such as flat belt, grooved belt, V-belt. See Figure 3-17 for examples.

(1) Flat belt. The flat belt is the simplest type that is often made from leather or rubber-coated fabric. The sheave surface is also flat and smooth, and the driving force is therefore limited by the pure friction between the belt and the sheave. Some designers prefer flat belts for delicate machinery because the belt will slip if the torque tends to rise too high to damage the machine.

(2) Synchronous belt. Synchronous belts, also called timing belts (Figure 3-17a), ride on sprockets having mating grooves into which the teeth are on the belt seat. This is a positive drive, limited only by the tensile strength of the belt and the shear strength of the teeth.

(3) Grooved belt. Grooved belts (Figure 3-17b) are a type of power transmission belt with grooves on the inner side. These belts are designed to operate in grooved pulleys and are commonly used in automotive engines and accessory drives. Grooved belts offer several advantages over traditional V-belts, such as increased surface contact with the pulley, improved

power transmission efficiency, and reduced slippage.

(4) V-belt. A widely used type of belt, particularly in industrial drives and vehicular applications, is the V-belt drive(Figure 3-17c). The V-shape causes the belt to wedge tightly into the groove, increasing friction and allowing high torques to be transmitted before slipping occurs. Most belts have high-strength cords positioned at the pitch diameter of the belt cross-section to increase the tensile strength of the belt. The cords, made from natural fibers, synthetic strands, or steel, are embedded in a firm rubber compound to provide the flexibility needed to allow the belt to pass around the sheave. Often an outer fabric cover is added to give the belt good durability.

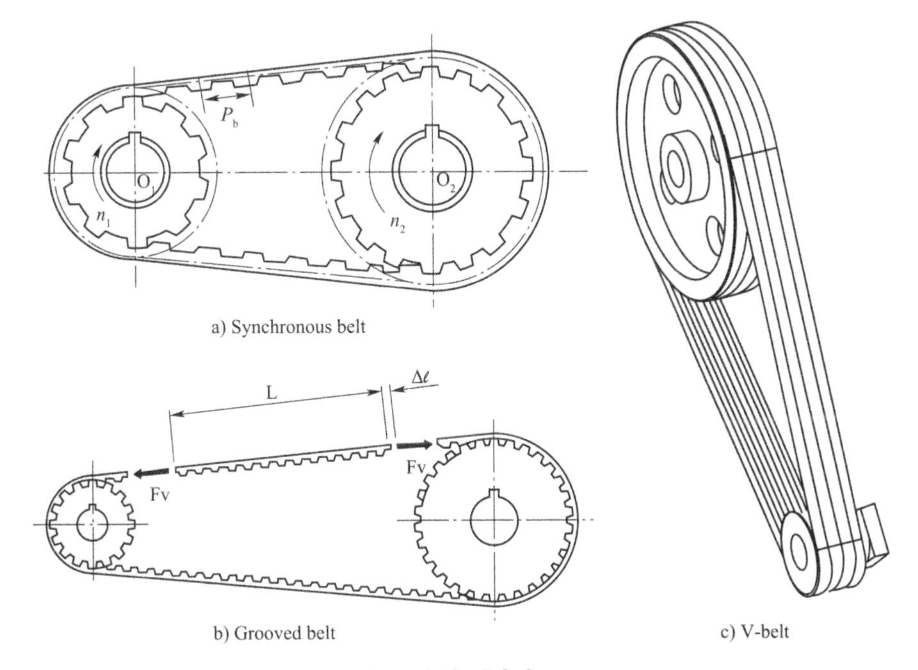

a) Synchronous belt

b) Grooved belt

c) V-belt

Figure 3-17 Belt drive

3) Design of belt drives

Designing belt drives involves selecting appropriate belts and pulleys to transmit power efficiently and reliably. Belt drives are widely used in various applications, from simple machines to complex systems. Here are the critical steps in the design process.

(1) Determine power requirements. Calculate the required power transmission based on the input and output speeds and the torque of the driven load.

(2) Select the belt type. There are several types of belts available, such as V-belts, flat belts, timing belts, and grooved belts. Choose the one that best suits the application based on load capacity, speed, and environmental conditions.

(3) Calculate belt length. Determine the required belt length based on the center distance between the pulleys and their diameters. Various online calculators or belt manufacturer catalogs can help with this calculation.

(4) Determine the number of teeth on timing belts (if applicable). Calculate the number of teeth on the pulleys for timing belt drives to ensure the correct gear ratio and synchronization between shafts.

(5) Select the pulley sizes. The pulley sizes affect the speed ratio and overall performance of the drive. Larger pulleys generally provide smoother operation and better load-carrying capacity.

(6) Check pulley dimensions. Based on the selected belt and the desired speed ratio, calculate the pulley dimensions to maintain the appropriate belt tension and avoid excessive wear.

(7) Verify pulley tooth profile (if applicable). For timing belt drives, ensure that the pulley tooth profile matches the belt profile to achieve proper engagement and avoid slippage.

(8) Consider environmental factors. Account for factors like temperature, dust, and corrosive substances affecting belts performance. Choose appropriate materials and coatings to enhance belt life.

(9) Verify safety factors. Ensure the selected belts and pulleys have adequate safety margins to handle peak loads and avoid premature failure.

3.5.2 Chain Drives

1) Structural characteristics of chain drives

As with belts, chain drives are used to transmit rotational motion and torque from one shaft to another smoothly, quietly, and inexpensively. Chain drives provide a belt drive's flexibility with a gear drive's positive engagement feature. Therefore, chain drives are well suited for applications with large distances between the respective shafts, slow speed, and high torque. Figure 3-18 shows a typical chain drive.

Figure 3-18 Chain drive

Compared to other forms of power transmission, chain drives have the following advantages:

(1) They are less expensive than gear drives.

(2) They have no slippage, as with belts, and provide more efficient power transmission.

(3) They have flexible shaft center distances, whereas gear drives are restricted.

(4) They are more effective at lower speeds than belts.

(5) They have lower loads on the shaft bearing because initial tension is not required as with belts.

(6) They have a longer service life and do not deteriorate with factors such as heat, oil, or age, as do belts.

(7) They require little adjustment, whereas belts require frequent adjustment.

2) roller chain

The most common type of chain is the roller chain, in which the roller on each pin provides exceptionally low friction between the chain and the sprockets.

The average tensile strengths of the various chain sizes are listed in Table 3-2. These data can be used for very low-speed drives or for applications in which the function of the chain is to apply a tensile force or to support a load. It is recommended that only 10% of the average tensile strength be used in such applications. For power transmission, a given chain size rating as a function of the rotation speed must be determined.

Roller chain sizes Table 3-2

Chain number	Pitch (in)	Roller diameter (in)	Roller width (in)	Plate thickness (in)	Average tensile strength (lb)
35	$\frac{3}{8}$	0.200	0.188	0.050	2100
40	$\frac{1}{2}$	0.312	0.312	0.058	3700
41	$\frac{1}{2}$	0.306	0.250	0.050	2500
50	$\frac{5}{8}$	0.400	0.375	0.079	6100
60	$\frac{3}{4}$	0.469	0.500	0.093	8500
80	1	0.625	0.625	0.125	14500
100	$1\frac{1}{4}$	0.750	0.750	0.157	24000
120	$1\frac{1}{2}$	0.875	1.000	0.189	34000
140	$1\frac{3}{4}$	1.000	1.000	0.219	46000
160	2	1.125	1.250	0.255	58000
180	$2\frac{1}{4}$	1.406	1.406	0.283	80000

Continue

Chain number	Pitch (in)	Roller diameter (in)	Roller width (in)	Plate thickness (in)	Average tensile strength (lb)
200	$2\frac{1}{2}$	1.562	1.500	0.312	95000
240	3	1.875	1.875	0.375	130000

3) Design of Chain Drives

Designing chain drives involves selecting appropriate components and dimensions to ensure efficient power transmission and reliable operation. Chain drives consist of two main components: the sprockets (or gears) and the chain. Here are the key steps in the design process.

(1) Determine power requirements. Calculate the required power transmission based on the input and output speeds and the torque of the driven load.

(2) Select chain type. There are various types of chains available, such as roller chains, silent chains, and inverted-tooth chains. Choose the one that best suits the application based on load capacity, speed, and environmental conditions.

(3) Calculate chain pitch. The chain pitch is the distance between the centers of two consecutive rollers. It is essential to select the correct pitch to ensure smooth engagement with the sprockets. The chain pitch is generally standardized for different chain sizes.

(4) Determine the number of teeth on the sprockets. The sprocket tooth count affects the gear ratio and overall performance. For a given gear ratio, the larger the sprocket, the smoother the operation and the higher the chain's capacity to carry loads.

(5) Calculate the center distance. The center distance is the distance between the centers of the two sprockets. It is crucial to maintain the proper center distance to avoid chain tension issues and excessive wear.

(6) Check chain tension. Ensure the chain tension is within acceptable limits to prevent excessive slack or overloading the chain.

(7) Calculate sprocket dimensions. Based on the chain pitch and the required tooth count, calculate the sprocket dimensions to ensure a proper fit with the chain and maintain the center distance.

(8) Check sprocket tooth profile. Ensure that the sprocket tooth profile is compatible with the selected chain type to achieve smooth engagement and minimize wear.

(9) Consider environmental factors. Account for factors like temperature, dust, and corrosive substances affecting chain performance. Choose appropriate materials and lubrication to enhance chain life.

(10) Verify safety factors. Ensure that the selected chain and sprockets have adequate

safety margins to handle peak loads and avoid premature failure.

4) Lubrication

Adequate lubrication must be provided for chain drives. There are numerous moving parts within the chain, along with the interaction between the chain and the sprocket teeth. The designer must define the lubricant properties and the method of lubrication.

(1) Lubricant Properties. Petroleum-based lubricating oil similar to engine oil is recommended. Its viscosity must enable the oil to flow readily between chain surfaces that move relative to each other while providing adequate lubrication action. The oil should be kept clean and free of moisture.

(2) Method of Lubrication. The American Chain Association recommends three different types of lubrication depending on the speed of operation and the power being transmitted.

① Manual or drip lubrication. For manual lubrication, oil is applied copiously with a brush or a spout at least once every 8h of operation. For drip feed lubrication, oil is fed directly onto the link plates of each chain strand.

② Bath or disc lubrication. The chain cover provides a sump of oil into which the chain dips continuously. Alternatively, a disc or a slinger can be attached to one of the shafts to lift oil to a trough above the lower strand of the chain. The trough then delivers a stream of oil to the chain. The chain itself, then, does not need to dip into the oil.

③ Oil stream lubrication. An oil pump delivers a continuous stream of oil on the lower part of the chain.

Chapter 4 Mechanical Manufacturing Technology

4.1 Concept and Organization Process of Mechanical Manufacturing

Manufacturing is the industry that processes various raw materials into usable industrial finished products. It not only provides technical equipment for various sectors of the national economy, but also provides material wealth for society. According to the statistics, 68% of the wealth of the United States comes from the manufacturing industry, 49% of Japan's gross national product is provided by the manufacturing industry. China's manufacturing industry also accounts for 40% of the gross national product. Thus, the machinery manufacturing industry is an important part of the manufacturing industry, an important foundation of the national industrial system. The improvement and progress of machinery manufacturing technology have a direct and important impact on the development of the entire national economy and the improvement of science, technology, and national defense strength. It is one of the important signs to measure a country's scientific and technological level and comprehensive national strength.

In a general sense, mechanical manufacturing refers to using blanks (or materials) and other auxiliary materials as raw materials, inputting them into the mechanical system, and finally, outputting qualified parts or products from the system through storage, transportation, processing, inspection, and other links. To sum up, machinery manufacturing is the sum of all kinds of labor that transform raw materials into finished products. The process generally includes the following stages.

(1) Technical preparation stage.

Before a certain part or product is put into production, among the technical preparations that must be done, the process procedure shall be formulated first, which is an important document to guide various technical operations. In addition, the supply of raw materials, the allocation of tools, fixtures, measuring tools, and the preparation of heat treatment equipment and testing instruments should be arranged in the technical preparation stage.

(2) Blank manufacturing stage.

The blanks can be obtained by different methods. The common methods for obtaining blanks are casting, forging, welding, and profile. Different blank-forming methods shall be selected according to the batch size, shape, performance requirements, and other factors.

Reasonable selection of blanks can increase productivity and reduce costs.

(3) Part processing stage.

Metal cutting is the main processing method for various parts at present. General processing equipment includes lathes, washing machines, drilling machines, planers, milling machines, grinders, etc. In addition, there are special machine tools, special processing machines, CNC machine tools, etc. The processing method and equipment to be used shall be comprehensively considered according to many factors, such as parts batch, precision, surface roughness, and various technical requirements, so as to ensure both parts quality requirements and high production efficiency and low cost.

(4) Product inspection and assembly.

Each part has a certain precision, surface roughness, and related technical requirements due to its different role in the machine. In the process of processing, it will inevitably produce processing errors. Therefore, the inspection procedure must be set to inspect the size and geometric shape errors generated during the processing. In addition, internal performance inspection, such as defect inspection, mechanical property, or metallographic structure inspection, shall also be carried out for parts working under heavy load, high temperature, and high pressure. Parts can only be used after they are fully qualified in quality inspection.

During the assembly process, the provisions of the technical conditions must be strictly observed, such as the cleaning of parts, assembly sequence, assembly method, tool use, joint surface grinding, lubricant application and running in, paint color and packaging. Only in this way can qualified products meet the requirements be produced.

4.2 Casting Technology and Its Application

Casting refers to the molding method of pouring liquid metal into a mold cavity that is suitable for the shape of the part, and obtaining a certain shape and performance of the part or blank after its solidification. The blanks or parts obtained by casting are called castings.

Casting has the following characteristics.

(1) It can be made into various castings with complex shapes, such as various boxes, beds, racks, etc.

(2) It has a wide range of applications. The metal materials commonly used in industry can be made into parts by casting. The weight of castings can range from a few grams to hundreds of tons, and the size can range from a few millimeters to tens of meters.

(3) Raw materials come from a wide range of sources and can be directly used for scrapped parts, chips, and scrap steel. Generally, expensive equipment is not required for casting, and the production cost of castings is low.

(4) The shape and size of the casting are close to the parts, so the workload of cutting is small, which can save metal materials.

The disadvantage of casting production is that liquid forming will bring some defects to the castings, such as loose casting structure, coarse grains, shrinkage cavity, porosity, slag inclusion, which will make the mechanical properties of castings lower than forgings of the same material. In addition, there are many casting processes and procedures, and the quality control factors are relatively complex. The working conditions of casting are poor.

Although casting has several shortcomings, its advantages are also obvious, so, it is widely used in industrial production. According to statistics, in metal cutting machine tools, the weight of castings accounts for 70% ~ 80% of the total weight; 45% ~ 70% of the total weight of automobiles and tractors; In some heavy machinery, it accounts for more than 85% of the total weight. With the development of casting technology, castings will be more and more widely used in all aspects of modern life.

There are many methods of casting production, including sand casting, metal mold casting, die casting, centrifugal casting, investment casting, etc. The most basic and common casting method is sand casting.

4.2.1 Casting Property of The Alloy

The performance of the alloy to obtain high quality castings in the casting process is called the casting property of the alloy. It is very important to understand the casting properties of alloys and their influencing factors for obtaining high-quality castings. The casting properties of the alloy mainly include fluidity, shrinkage, oxidation, gas absorption, etc. Among them, fluidity and shrinkage have the greatest influence on the casting properties of the alloy.

1) Fluidity

(1) Influence of fluidity on casting quality.

The fluidity of liquid metal is called fluidity, and the fluidity of liquid metal affected by mold and process factors is called mold filling capacity. The fluidity directly affects the filling ability of liquid metal. The influence of fluidity on casting quality is shown in three aspects.

① The alloy with good fluidity can easily obtain castings with complete shape, accurate size, and clear contour. For thin-walled and complex castings, the quality of alloy fluidity is often the decisive factor in obtaining qualified castings. The alloy with poor fluidity is easy to cause defects such as cold shut and insufficient pouring.

② In liquid alloys, there are often a certain amount of gas and non-metallic inclusions. The liquid alloy with good fluidity makes it easy to let gas escape before and during pouring and blocks non-metallic inclusions floating on the liquid surface, which ensures the internal quality of the casting. The liquid alloy with poor fluidity can easily produce slag inclusion, porosity, and other defects in the casting.

③ In the cooling and solidification process of castings, volume shrinkage will occur. Alloys with good fluidity can make the solidification shrinkage part of liquid alloy timely supplemented by liquid alloy, thus preventing casting from shrinkage cavity, porosity, and other defects.

(2) Factors affecting liquidity.

① Chemical composition. In the phase diagram of iron-carbon alloy, the alloy with eutectic composition has the best fluidity.

② Pouring conditions. The higher the pouring temperature of the alloy, the longer the liquid is kept, and the lower the viscosity of the liquid alloy, the stronger the filling ability of the liquid alloy. The higher the pressure and flow rate of the liquid alloy during pouring, the more conducive to filling the mold.

In production, for thin walls, complex shapes, and poor fluidity alloys, measures such as increasing pouring temperature, increasing pressure of liquid alloy, and increasing pouring speed are often taken to improve the filling ability of liquid alloy. For example, adding a sprue cup to the upper mold is equivalent to heightening the sprue, which can increase the pressure of liquid alloy. Increasing the cross-section size of the gating system can improve the pouring speed of liquid alloy, but the casting easily produces sand, porosity, shrinkage cavity, and other defects when the pouring temperature is too high. Increasing the cross-sectional size of the gating system will increase the consumption of liquid alloy, etc. It can be seen that in the process of casting process design, comprehensive consideration should be given to the reasonable selection of pouring parameters.

③ Mold material and mold structure. The influence of the mold on the filling ability of liquid metal is mainly shown in the resistance and thermal conductivity of the mold to the flow of liquid metal. The more complex the shape of the casting and the smaller the wall thickness, the greater the resistance when the liquid alloy flows, and the faster the temperature of the liquid alloy decreases, which will inevitably reduce the filling ability of the liquid alloy. In production, a metal mold is more likely to cause casting defects such as insufficient pouring and cold shut than sand mold, and wet mold is more likely to cause casting defects than dry mold. The reason is that the former has a strong mold thermal conductivity, and the temperature of the liquid alloy drops rapidly, thus reducing the filling capacity of the liquid alloy.

2) Shrinkage

In the process of liquid solidification and cooling to room temperature, the volume and size of the alloy decrease, which is called shrinkage. It includes three stages: liquid shrinkage, solidification shrinkage, and solid shrinkage.

(1) Influence of shrinkage on casting quality.

① Liguid shrinkage. The volume shrinkage of liquid metal due to temperature reduction is called liquid shrinkage.

② Solidification shrinkage. The volume shrinkage of liquid metal during solidification is called solidification shrinkage. The solidification shrinkage of pure metals and alloys crystallized at constant temperature is simply caused by the liquid-solid phase transition. For alloys with a certain range of crystallization temperatures, in addition to the shrinkage caused by the liquid-solid phase transformation, there is also the shrinkage caused by the temperature drop in the solidification stage.

③ Solid shrinkage. The volume shrinkage of metal in the solid state due to temperature reduction is called solid shrinkage.

The volume change of alloy caused by liquid shrinkage and solidification shrinkage is called "volume shrinkage", which is the main reason for the shrinkage cavity and porosity of castings. Although solid shrinkage can also cause volume change, it is mainly manifested in three aspects of casting: shrinkage of outlet size, which is called "linear shrinkage", which is the main reason for residual internal stress, deformation, and cracking of castings.

(2) Factors affecting contractility.

① Chemical composition. Graphite is precipitated from the gray cast iron during crystallization. As the specific volume of graphite is large and the density is small, the precipitation of graphite compensates for part of the shrinkage of the cast iron. Carbon and silicon are elements that promote graphitization in cast iron. Therefore, with the increase of carbon and silicon content, the shrinkage of cast iron decreases. Sulfur is an element that hinders graphitization, so the greater the sulfur content, the greater the shrinkage of gray cast iron.

② Pouring temperature. The higher the pouring temperature, the greater the liquid shrinkage, and therefore the greater the volume shrinkage.

③ Mold material and casting structure. The mold cavity and core block the bulk shrinkage of the alloy. In addition, because the wall thickness of the casting cannot be uniform, the speed of alloy solidification and cooling cannot be the same everywhere. The part that solidifies first and cools down restricts the shrinkage of the part that solidifies later and cools down. The above-mentioned blocking and holding effects can reduce the linear shrinkage of the alloy in the solid state.

(3) Effect of alloy shrinkage on casting quality.

① Shrinkage cavity and porosity in the casting.

When the liquid alloy is cooled and solidified in the casting mold, the volume will decrease due to the liquid shrinkage and solidification shrinkage. If the alloy liquid is not supplemented, holes will be formed in the last solidified part of the casting. The large and concentrated holes are called shrinkage cavities; The small and concentrated relict L holes are called shrinkage porosity. The shrinkage cavity porosity is a casting defect that can not be ignored.

Due to the narrow crystallization temperature range, the solidification of pure metal and near eutectic alloy is carried out layer by layer from the outside to the inside, which makes it easy to form concentrated shrinkage cavities. For alloys with a large crystallization temperature range, the developed dendritic crystals will separate the uncured liquid metal, which can easily form shrinkage porosity. Shrinkage cavity and porosity will affect the mechanical properties, air tightness, and physical and chemical properties of castings. The method to prevent shrinkage cavity and porosity in production is usually using a riser, chill, etc., to achieve directional solidification of castings and supplement the shrinkage of metal liquid volume. It is difficult to prevent shrinkage porosity. For castings requiring high air tightness, attention should be paid to selecting alloys with a small crystallization temperature range.

② Casting stress, deformation, and crack.

After solidification, the casting will begin to contract in the solid state during the process of continuous cooling. If the shrinkage is blocked, stress will be generated in the casting, which is called casting stress. Casting stress is the main reason for the deformation and crack of castings.

The casting stress can be divided into thermal stress, shrinkage stress, and phase transformation stress according to different causes. Thermal stress is the stress caused by uneven shrinkage in different parts of castings during solidification and cooling. When the casting is cooled, the greater the temperature difference and the shrinkage of the alloy, the greater the thermal stress formed. Shrinkage stress is the stress caused by external forces such as mold, core, gate, and riser when the casting shrinks in the solid state. Phase transformation stress is the stress caused by the uneven volume change of each part of the casting due to solid phase transformation.

When the stress in the casting reaches a certain size, the casting will be deformed and cracked. In order to prevent deformation and cracks of castings, the casting stress of castings shall be reduced. For example, for ladder-shaped castings, the temperature of each part of the casting tends to be uniform by adjusting the position of the ingate, placing chill, and other measures to achieve simultaneous solidification of each part of the casting and greatly reduce the thermal stress. The existing casting stress on the casting shall be eliminated by heat treatment or natural aging.

4.2.2 Casting Process and Method

There are many methods of casting production, including sand casting, metal mold casting, die casting, centrifugal casting, investment casting, etc. The most basic and common casting method is sand casting.

The method of compacting and molding with molding sand is called sand casting. The castings produced by sand casting account for more than 90% of the total weight of all castings. Figure 4-1 shows the flow chart of the sand casting process.

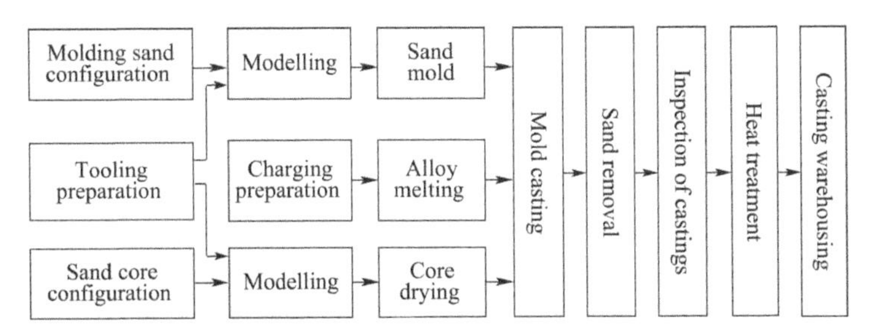

Figure 4-1 Sand casting process flow chart

It can be seen from the figure that the sand casting production process includes pattern making, molding material preparation, modeling, core making, mold fitting, smelting, pouring, vacuum casting, cleaning and inspection, etc. Among them, modeling and core making are important links to sand casting, which have a great impact on the quality of castings.

other casting methods different from sand casting are called special casting. With the development of casting technology, special casting plays an important role in casting production. Under specific conditions, special casting can improve the dimensional accuracy of castings, reduce surface roughness, improve metal properties, increase productivity, and improve working conditions. Common special casting methods include metal mold casting, pressure casting, centrifugal casting, investment casting, low pressure casting, ceramic mold casting, continuous casting and squeeze casting.

Here the procedures of sand casting will be elaborated.

1) Molding materials

The materials used to make casting mold are called molding materials. The material used to make sand mold is called molding sand. The material used to make the core is called core sand.

(1) Requirements for properties of molding sand and core sand.

The indexes required for the properties of molding sand and core sand include strength, air permeability, fire resistance, deformability, plasticity, etc.

① Strength. Refers to the ability of molding sand and core sand to withstand external forces without being damaged after molding. The molding sand and core sand shall have enough strength to prevent damage, kick and swell during handling, overturning, boxing, and pouring metal. If the strength of molding sand and core sand is not good, castings are prone to sand holes, sand inclusions, and other defects.

② Air permeability. Refers to the ability of molding sand and core sand to penetrate gas. During the pouring process, when the casting mold contacts with high-temperature metal liquid, the gas released from water vaporization, organic matter combustion, and liquid metal cooling must be discharged through the casting mold otherwise air holes will be generated in the casting or the casting will be insufficiently poured.

③ Fire resistance. Refers to the ability of molding sand and core sand to withstand high-temperature heat. The refractoriness mainly depends on the content of SiO_2 in quartz sand. If the fire resistance is not enough, a layer of sticky sand will be formed on the surface of the casting or in the inner cavity, which not only makes it difficult to clean, affects the appearance, but also increases the difficulty of machining.

④ Deformability. Refers to the property that molding sand and core sand can be compressed and conceded when shrinkage occurs during solidification and cooling of castings. The insufficient yield of molding sand and core sand will hinder the shrinkage of castings, resulting in internal stress, deformation, cracks, and other defects.

⑤ Plasticity. Refers to the ability of molding sand, and core sand to deform under external force and retain deformation after removing external force. With good plasticity, molding sand and core sand are soft and easy to deform and are not easy to break and drop during mold lifting and repair.

In addition to the above performance requirements, there are also performance requirements such as collapsibility, gas generation, and moisture absorption. Many properties of molding sand and core sand are sometimes contradictory. For example, high strength and good plasticity may reduce permeability. Therefore, the specific proportion of molding sand and core sand should be determined according to the type of casting alloy, casting size, structure, etc.

(2) Composition of molding sand and core sand.

Molding sand and core sand are mainly composed of raw sand, binder, additives, coatings, and fillers.

① Raw sand. The main component of raw sand is silica sand, while the main component of silica sand is SiO_2, and its melting point is up to 1700℃. The higher the SiO_2 content in the sand, the higher the refractoriness; The coarser the sand, the higher the fire resistance and air permeability; The silica sands with polygonal and angular shapes have good air permeability; The smaller the mud content is, the better the air permeability is.

② Binder. The material used to bond sand particles is called binder. The commonly used binders are clay and special binders. Its electric clay is the main binder for preparing molding sand and core sand, which is divided into bentonite and ordinary clay. Bentonite with good binder performance is widely used as green sand. Common clay is often used for dry molding sand. Special adhesives include tung oil, water glass, resin, etc. These special binders are often used for core sand.

③ Additives. The materials added to improve some properties of molding sand and core sand are called additives. For example, adding pulverized coal can reduce the roughness of the casting surface and inner cavity; Adding wood chips can improve the flexibility and permeability of molding sand and core sand.

④ Coatings and rolling materials. These materials are not the ingredients added when preparing molding sand and core sand, but are used to spread (dry) or disperse (wet) on the surface of the mold to reduce the roughness of the casting surface to prevent the occurrence of sand sticking defects.

2) Modeling method

The process of using molding sand, patterns, and other technological equipment to manufacture molds is called molding. Modeling methods are generally divided into manual modeling and machine modeling.

(1) Manual modeling.

All modeling methods completed by hand or hand tools are called manual modeling. Manual modeling is characterized by flexible operation, strong adaptability, low pattern cost, and simple production preparation, but low modeling efficiency, high labor intensity, and poor labor environment. It is mainly used for single-piece and small-batch production. How to take the pattern out of the sand mold smoothly without damaging the shape of the mold cavity is the key problem. Therefore, centering on the problem of how to lift the mold, the modeling methods are divided into full mold modeling, split mold modeling, sand digging modeling, fake box modeling, live block modeling, three box modeling, and scraper modeling.

(2) Machine modeling.

The molding method of completely completing or at least completing the sand compaction operation with a machine is called machine modeling. The essence of machine modeling is that machines replace manual sand tightening and mold lifting. Machine modeling shall be adopted for mass production. Compared with manual molding, machine molding has high production efficiency, high dimensional accuracy of castings, and good surface quality, but it requires high equipment and process equipment, and takes a long time to prepare for production.

3) Core making

The process of core making is called core making. The main function of the core making is to obtain the inner cavity of the casting, but sometimes it can also be used as the local mold of the casting that is difficult to lift the mold. During pouring, core sand shall have higher strength, air permeability, and fire resistance than molding sand due to impact, surrounding and baking of molten metal. In order to meet the above performance, the following measures should be taken.

(1) Open air hole and vent.

The core with simple shape can pierce the vent hole with a vent needle. For the core with complex shape, wax wire or straw rope can be put into the core. When drying, wax wires or straw ropes are burned to form an air passage, so as to improve the ventilation of the core, as shown in Figure 4-2a) and b).

(2) Core bone and installation ring.

The core bone is a metal frame placed in the sand core to strengthen or support the sand core. For larger cores, the core bone and lifting ring are often placed in the core to improve the strength of the core and faciliate lifting, as shown in Figure 4-2c). The small core bone is generally made of iron wire, and the large core bone with complex shape is cast from cast iron.

The core can be made by hand or machine. The core box is mainly used for manual core making. When single piece and small batch production of large and medium-sized rotary cores, the scraper can be used for core making. Among them, core box making (as shown in Figure 4-3) is the most commonly used method, which can produce cores with complex shapes.

4) Gating system

A series of channels for liquid metal to flow into the mold cavity are called gating systems. The function of the gating system is to ensure that the liquid metal flows into and fills the cavity evenly and stably, so as to avoid damaging the cavity; prevent slag, sand or other impurities from entering the mold cavity; adjust the solidification sequence of castings or replenish liquid metal required for condensation and contraction of metal liquid. The gating system is an important part of the casting mold. If the design is not reasonable, the casting is prone to sand washing, sand holes, insufficient pouring, and other defects. A typical gating system consists of the following parts, as shown in Figure 4-4.

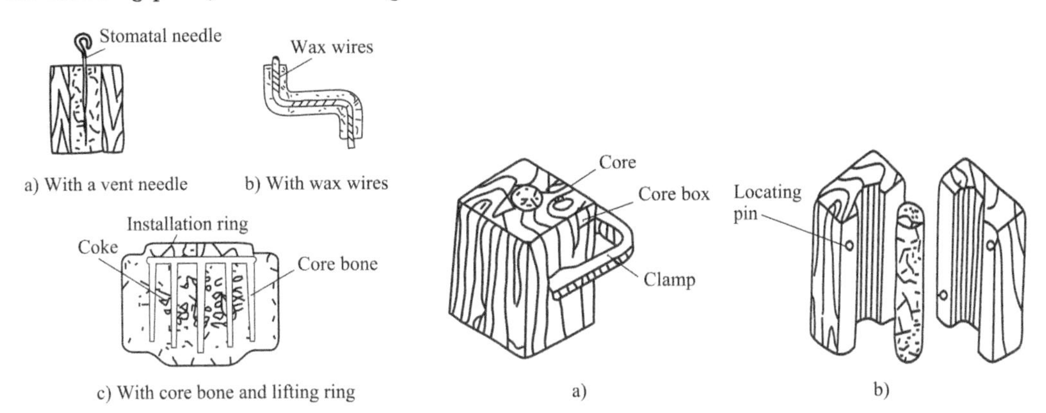

Figure 4-2 Core making measures

Figure 4-3 Schematic diagram of core box coremaking

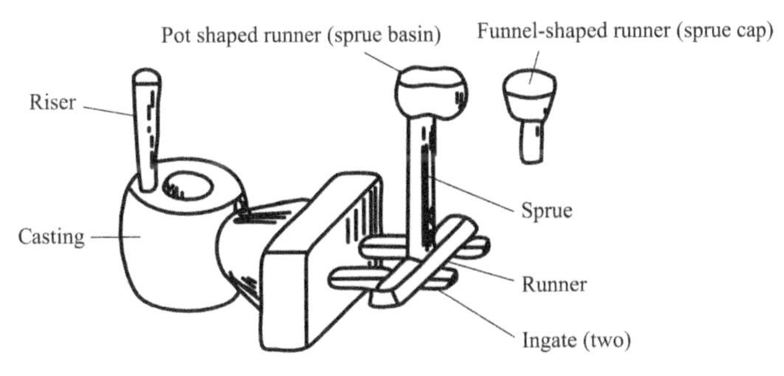

Figure 4-4 Gating system of castings

The role of the outer runner is to ease the middle force of the liquid metal and make it flow smoothly into the sprue. The sprue is a section of the upper large and lower small conical channel under the outer runner. Its certain height makes the liquid metal produce a certain static pressure, so that the metal liquid can fill the cavity with a certain flow rate and pressure. The runner is located above the ingate and is a trapezoidal channel with small upper part and large lower part. As the runner is higher than the ingate, the slag and sand particles in the liquid metal will float on the top of the runner to prevent slag inclusion and sand inclusion. In addition, the runner also plays the role of distributing molten metal to the ingate. The section of the ingate is mostly flat trapezoid, which plays a role in controlling the flow direction and velocity of liquid metal. The function of the riser is to supplement liquid metal during solidification shrinkage of liquid metal to prevent shrinkage defects of castings. In addition, the riser also plays the role of exhaust and slag collection. The riser is generally located at the highest and thickest part of the casting.

5) Mold closing, smelting, and pouring

(1) Mold closing.

The process of forming a complete mold from each component of the mold (upper mold, lower mold, sand core, gating basin, etc.) is called mold closing. During mold closing, check whether the mold cavity is clean, whether the core is installed accurately and firmly, and whether the sandbox is positioned accurately and firmly.

(2) Smelting.

The operation process of changing metal from solid to liquid by heating, removing impurities in metal by metallurgical reaction, and making its temperature and composition meet the specified requirements is called smelting. If the temperature of molten metal is too low, the casting will have defects such as cold shut, insufficient pouring, and air hole. If the temperature of the molten metal is too high, it will lead to an increase of the total shrinkage of the casting, excessive absorption of gas, sand, and other defects. The smelting equipment commonly used in foundry production includes cupola (for smelting cast iron), electric arc furnace (for smelting cast steel), Xunyin furnace (for smelting non-ferrous metals), and induction heating furnace (for smelting cast iron and cast steel).

(3) Pouring.

The operation process of injecting molten metal into the mold from the ladle is called pouring. The pouring temperature of cast iron is 20℃ above the liquidus (1250 ~ 1470℃ in general). If the pouring temperature is too high, the liquid metal will absorb more gas, shrink more liquid, and the casting will be prone to air holes, shrinkage cavities, sand, and other defects. If the pouring temperature is too low, the liquid metal will have poor fluidity, and the casting will be prone to defects such as insufficient pouring and cold shut.

6) Sand falling, cleaning, and inspection

(1) Sand falling.

The operation process of separating casting from molding sand (core sand) and sandbox manually or mechanically is called sand falling. After pouring, sand can only be removed after full solidification and cooling. If the sand falls too early, the cooling rate of the casting is too fast, which will cause the white structure on the surface of the cast iron, resulting in difficult cutting. If the sand falls too late, the casting will crack due to the large shrinkage stress, and the productivity will be low.

(2) Clearing.

After sand falls, the operation process of removing surface sand, molding sand (core sand) by mechanical cutting and hammering and removing excess metal (gate, riser, fins and oxide skin) by gas cutting is called cleaning.

(3) Inspection.

The quality inspection shall be carried out after the castings are cleaned. The surface defects of castings can be found by eye observation (or with the help of a pointed hammer), such as air holes, sand holes, sticky sand, shrinkage cavities, under pouring and cold shut. For internal defects of castings, pressure test, and ultrasonic flaw detection can be carried out.

4.2.3 Structural Processability of Castings

1) Requirements of alloy casting performance on casting structure

(1) The wall thickness of the casting shall be reasonable.

The greater the wall thickness of the casting is, the smaller the resistance of the liquid metal when flowing, and the longer the liquid metal is kept, so it is beneficial for the liquid metal to fill the cavity. However, with the increase of the wall thickness, the cooling rate of the molten metal decreases, and coarse grains are easily obtained in the center of the casting, which will reduce the mechanical properties of the casting alloy. When the wall thickness of the casting is reduced, fine grains can be obtained and the mechanical properties of the casting can be improved. However, if the wall thickness of the casting is too small, its fluidity will deteriorate due to the rapid cooling of the molten metal, and it is easy to have cold shut and insufficient pouring on the casting Sink.

Generally speaking, the wall thickness of the casting should first meet the requirements of alloy fluidity, and then try not to make the wall thickness of the casting too large. The minimum thickness of the casting alloy that can fill the mold is called the minimum wall thickness of the casting alloy. The wall thickness should meet the requirement of minimum wall thickness.

(2) The wall thickness of castings shall be uniform.

If the wall thicknesses of castings vary too much, there will inevitably be slow cooling hot spots at the wall thicknesses, where shrinkage cavities, porosity, coarse grains, and other defects are easy to form. At the same time, due to the different cooling rates of different wall thicknesses, thermal stress will be generated between the thick wall and the thin wall, which may lead to hot cracks. As shown in Figure 4-5, the two casting structures on the left are examples of unreasonable wall thickness design, and the two casting structures on the right are the examples of reasonable wall thickness.

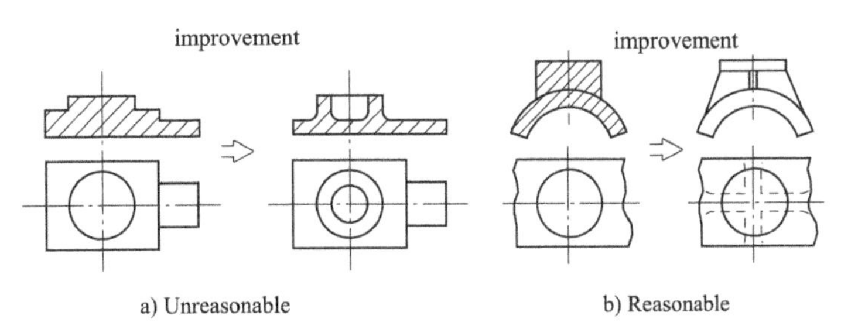

Figure 4-5 Example of wall thickness design

(3) The connection between walls should be reasonable.

The following three points should be noted for wall-to-wall connection.

Structural fillets are required. If the corner of the casting is connected at a right angle, not only will hot spots be formed here, which is easy to produce the crystal fragile zone of shrinkage joint, but also cracks will be generated at the crystal fragile zone caused by stress concentration, as shown in Figure 4-6.

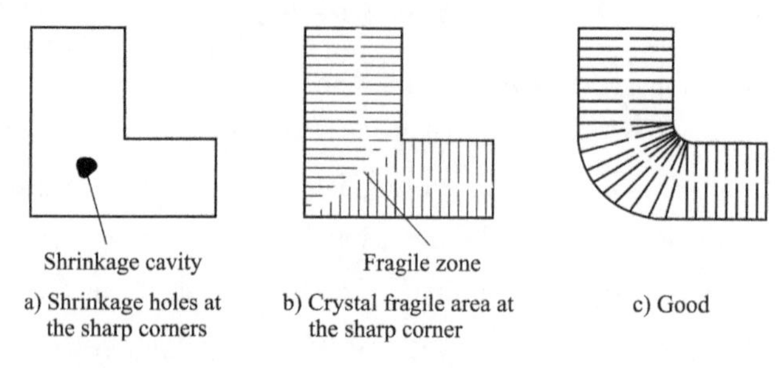

Figure 4-6 Effect of fillet and sharp corner on casting quality

The junction of wall thickness shall be transited reasonably. It is difficult for the wall thickness of each part of the casting to be completely consistent. At this time, care should be taken to avoid sudden changes in the connection between the thick wall and the thin wall, and make it transition gradually.

The connection between walls shall avoid crossing an acute angle. Hot spots are often formed at the junction of more than two casting walls. Shrinkage defects can be prevented if the cross structure and acute angle intersection can be avoided. Figure 4-7 shows the comparison of several connection structures.

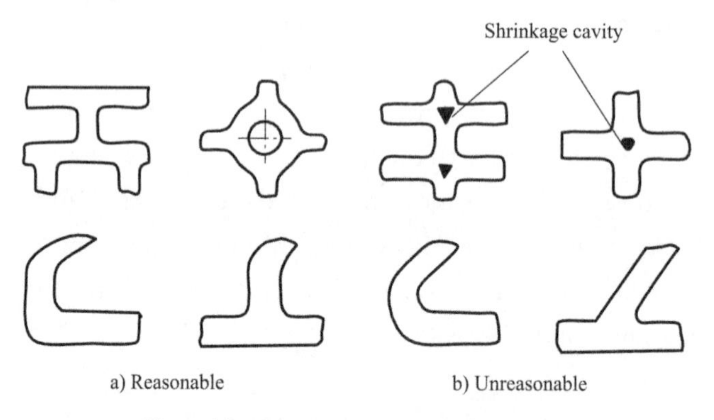

Figure 4-7 Connection structure between walls

(4) Convenient feeding shall be considered at the thick wall of the casting.

When there must be thick wall part in the casting, in order to avoid shrinkage cavity in the thick wall part, the structure of the casting shall have the conditions of sequential solidification and feeding. For the two castings in. Figure 4-8a), because the upper wall thickness is smaller than the lower wall thickness, the upper part solidifies faster than the lower part, so the feeding channel from top to bottom is blocked, and shrinkage cavities are easy to occur at the thick wall. If the structure is changed as shown in Figure 4-8b), the casting can be fed by the riser.

(5) The casting shall avoid large horizontal plane as much as possible.

The horizontal surface on the casting is not conducive to the filling of liquid metal. At the same time, the sand on the surface is also easy to drop, resulting in sand inclusion and other defects of the casting. Figure 4-9 shows the comparison scheme of the casting structure.

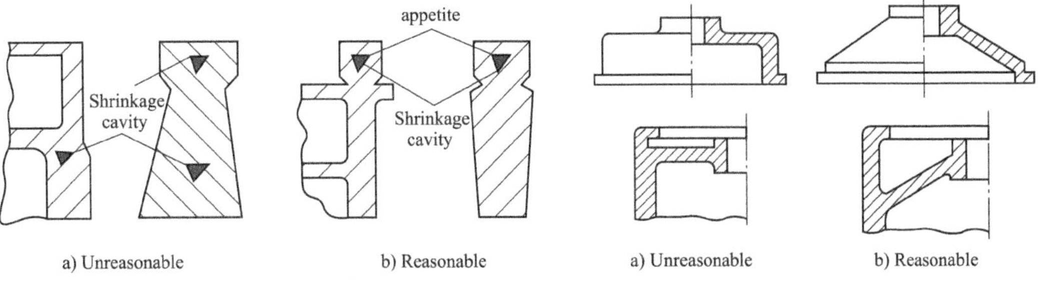

a) Unreasonable b) Reasonable

Figure 4-8 Casting structure considering feeding

a) Unreasonable b) Reasonable

Figure 4-9 Casting structure

(6) Avoid blocking during casting shrinkage.

In the final shrinkage part of the casting, if it cannot shrink freely, there will be tension. Due to the low tensile strength of the alloy at high temperatures, castings are prone to hot cracking defects. As shown in Figure 4-10, when the spokes of the wheel are straight and even, cracks are easily generated at the spokes. If the spokes are designed to be odd and curved, with help of which, the stress during contraction can be reduced, so as to avoid hot cracking.

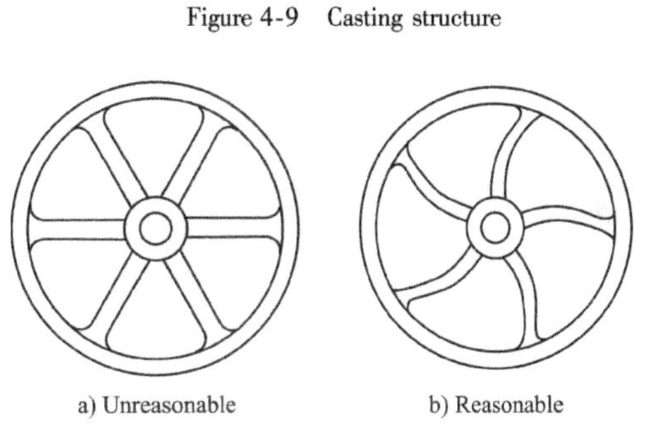

a) Unreasonable b) Reasonable

Figure 4-10 Spoke design

(7) Try to avoid reducing the bearing capacity due to openings on the wall.

Holes in the casting wall often cause stress concentration and reduce the bearing capacity. In case of necessity, in order to enhance the bearing capacity of the opening on the wall, a boss is generally set at the opening, as shown in Figure 4-11.

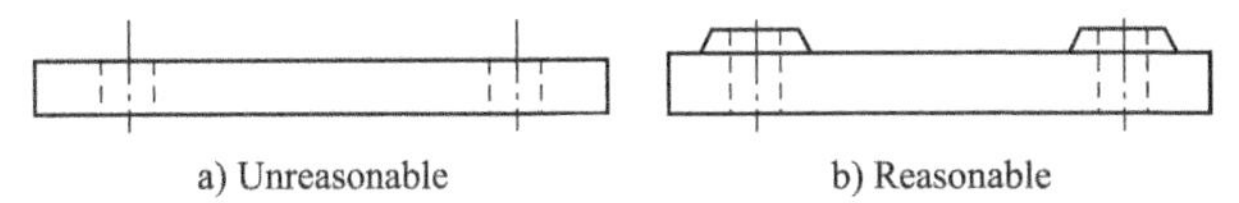

a) Unreasonable b) Reasonable

Figure 4-11 Boss to enhance the bearing capacity at the opening

Flat and slender castings often have warping or bending deformation due to uneven cooling. As shown in Figure 4-12a), the three castings are prone to deformation. The casting deformation can be effectively prevented by adding a reinforcing rib smaller than the plate thickness on the plate, or by changing the asymmetric structure to a symmetric structure, as shown in Figure 4-12b).

2) Requirements of casting process on casting structure

(1) Simplify the casting structure and reduce the parting surface.

The molding workload accounts for about one third of the total sand casting workload. Therefore, reducing the molding workload is an important measure to improve production efficiency. The less parting surface can reduce the use of sandbox and molding time, and also reduce the casting defects caused by wrong mold and eccentric core.

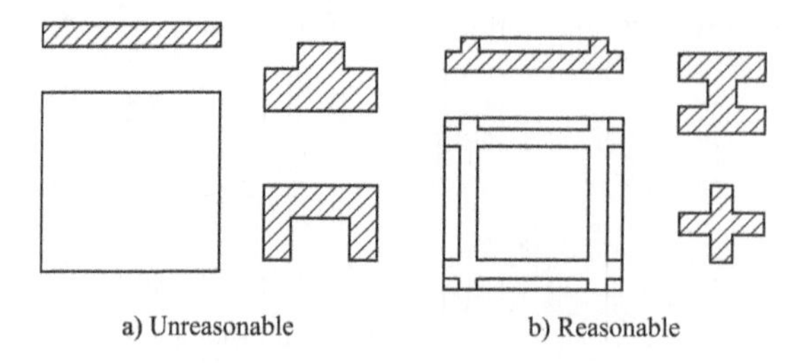

Figure 4-12 Casting structure to prevent deformation

(2) Try to use plane parting surface.

In the modeling process, the parting surface should be plane as much as possible, and unnecessary fillets should be removed. The plane parting surface can avoid sand digging and false box modeling, and has high productivity.

(3) Try to use fewer or no cores.

Reducing or not using cores can save the use of core making materials and drying cores, as well as reduce core making, core lowering, and other operations. For this reason, the mold cavity should be obtained by using naturally formed sand stacks (the upper mold is called suspended sand, and the lower mold is called self contained core) as much as possible.

(4) Try not to use or use fewer live blocks.

If there is a boss on the side wall of the casting, the movable block molding can be used. However, the movable block molding method requires a large amount of modeling work and is difficult to operate. If the boss not far from the parting surface is extended to a place convenient for mold lifting, the lifting block operation can be avoided or reduced.

(5) The structural slope of the vertical wall shall be considered.

If the non-machined surface perpendicular to the parting surface has a certain structural slope, it is not only convenient to lift the mold, but also improves the dimensional accuracy of the casting because the pattern does not need greater looseness.

(6) The setting of the core shall be stable and solid, and conducive to the exhaust and cleaning.

Only when the core is firmly fixed in the mold can the core deviation be avoided, only when the air outlet channel is unobstructed can air holes be avoided, and only when the sand is convenient for cleaning can the cleaning time be reduced.

4.3 Pressure Processing and Its Application

The forming method of applying external force to the blank to make plastic deformation, and change its size and shape is called forging, which is used to manufacture mechanical parts or blanks. It is the general term for forging and stamping. Forging methods mainly include free forging, mould forging, hammer die forging, stamping, rolling, extrusion, and drawing. Forging production is one of the main ways to obtain blanks in the machinery manufacturing industry. In machine tool manufacturing, the main shaft, transmission shaft, gears, and other important parts as well as cutting tools are formed by forging. The weight of forged parts on automobiles accounts for about 70% of the total weight of metal parts. Forging production has also been widely used in transportation, electric power, national defense, agriculture, and daily necessities.

Forging processing has the following advantages.

(1) The internal structure of the metal is improved, and the mechanical properties of the metal are improved. This is because forging can press the loose parts in the blank, and improve the density of metal. It can refine coarse grains. The carbides in the high alloy tool steel can be broken and evenly distributed.

(2) Save metal materials. Because forging improves the strength and other mechanical properties of metal, the section size of parts under the same load is relatively reduced, and the quality of parts is reduced. In addition, when precision forging is used, the dimensional accuracy and surface roughness of the forging parts can be close to the finished parts, so that there is little or no cutting.

(3) It has high productivity. The production efficiency of forging, especially die forging, is much higher than that of cutting. For example, in the production of socket head cap screws, the productivity of die forging is 50 times that of cutting. If the cold process is adopted, its production efficiency is more than 400 times that of cutting and forming.

(4) It has strong adaptability. Forging processing can not only manufacture forgings with simple shapes (such as round shafts), but also forgings with complex shapes that do not require or only require a small amount of machining (such as precision forging gears). The weight of forgings can be as small as less than one gram and as large as hundreds of tons. Forgings can also be produced in small batches or large quantities.

Disadvantages of forging production: The commonly used free forging precision is relatively low. The die cost of mould forging and die forging is high. Compared with casting production, producing blanks with complex shapes and cavities is difficult.

4.3.1 Principle of Pressure Processing

1) Basic principle of plastic deformation

(1) Plastic deformation of single crystal.

The plastic deformation of a single crystal is mainly realized by the slip form of the single crystal. That is, under the action of shear stress, one part of the crystal slides along a certain crystal plane relative to the other, as shown in Figure 4-13. The movement of a large number of dislocations in the crystal constitutes a macroscopic slip, as shown in

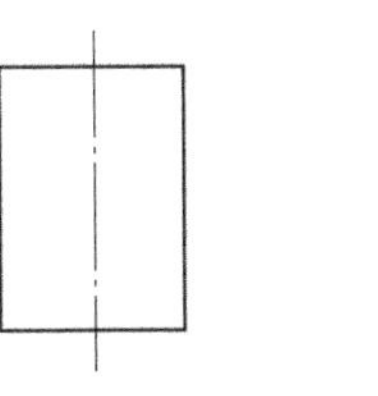
a) Before slip

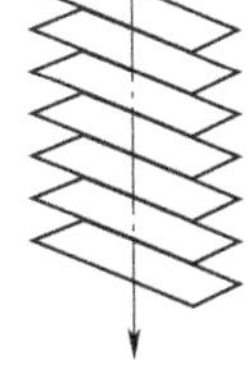
b) After slip deformation

Figure 4-13 Schematic diagram of crystal slip

Figure 4-14.

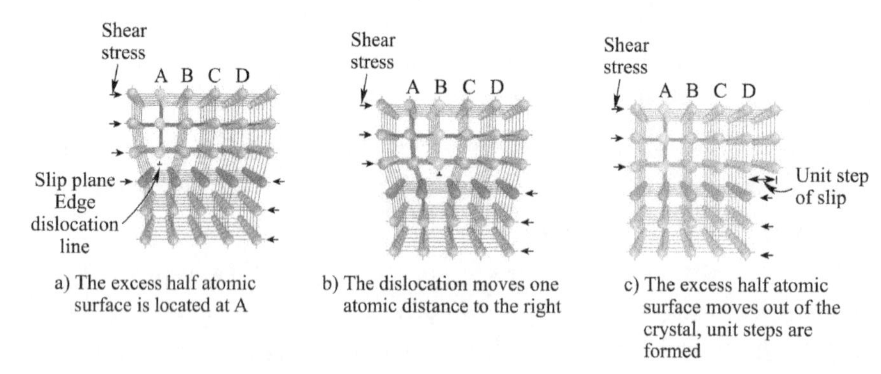

Figure 4-14 Diagram of dislocation movement

Twin deformation is another form of single crystal plastic deformation. When the shear stress acting on the crystal reaches a certain value, one part of the crystal is sheared relative to the other, and the crystal structure of the sheared part and the unshared part is symmetrically distributed, as shown in Figure 4-15 (note the part with hatched lines in the figure).

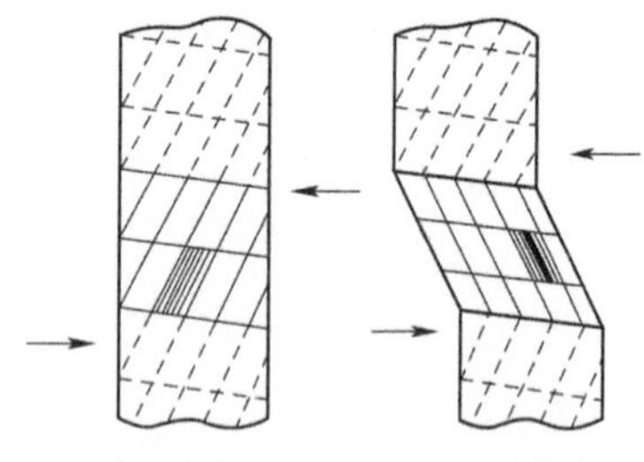

Figure 4-15 Schematic diagram of deformation caused by strain

(2) Plastic deformation of polycrystals.

Polycrystals are composed of many small grains with different orientations. Because the plastic deformation of each grain is restricted by the surrounding grains and crystal planes, the plastic deformation of polycrystals is much more complex than that of single crystals.

The plastic deformation of polycrystals can generally be summarized into two forms: the plastic deformation of grains themselves and the plastic deformation between grains.

2) Effect of Cold Deformation on Microstructure and Properties of Metals

When a metal is cold deformed (the deformation temperature is lower than the recrystallization temperature), many fine crystals (also known as broken crystals) will be generated on its slip plane and between grains. At the same time, the lattice near the deformation area will also produce distortion (also known as distortion).

During cold deformation, all strength indexes and hardness has increased, but the plasticity has decreased. This phenomenon is called cold deformation strengthening. Cold deformation strengthening is one of the important ways to strengthen metals.

3) Recovery and recrystallization of metals

When the cold deformation strengthening structure is heated, the deformed metal will undergo three stages of recovery, recrystallization, and grain growth, as shown in Figure 4-16.

(1) Recovery.

When the heating temperature is not high, the activity of the metal atom is not enough, but it can recover from the unstable position to the stable position, making the lattice distortion disappear and the internal stress reduced. At this time, the mechanical properties change little, the strength decreases slightly, and the plasticity rises slightly. This process is called recovery.

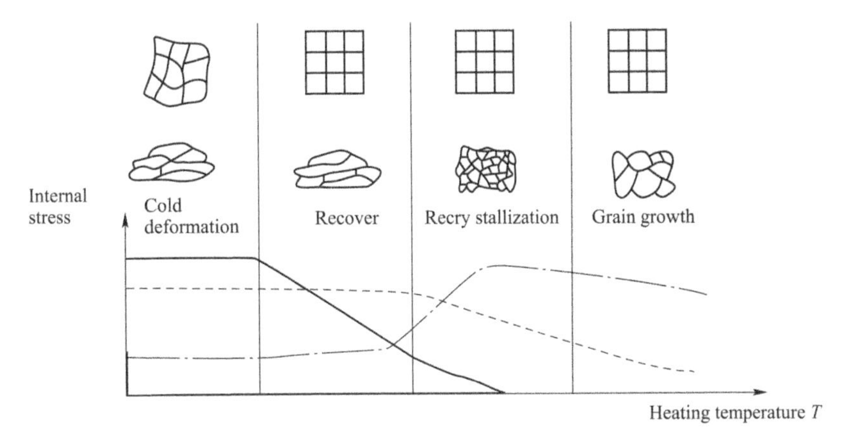

Figure 4-16 Effect of heating on microstructure and mechanical properties of cold deformation strengthened metals

(2) Recrystallization.

With the increase of heating temperature, the ability of atomic activity is enhanced. After cold deformation, the elongated grains of metal re-nucleate, crystallize and become equiaxed grains. At this time, the strength and hardness of the metal decrease while the plasticity increases significantly. This process is called recrystallization.

(3) Grain growth.

When the heating temperature exceeds the recrystallization temperature too much, the grains will grow up obviously and become coarse, and the grain structure deteriorates the malleability of the metal.

4) Hot working streamline and forging ratio

(1) Hot working streamline.

During forging, the brittle impurities of the metal are broken and distributed in granular or chain shape along the main elongation direction of the metal, while the plastic impurities are distributed in strip shape along the main elongation direction along the metal deformation so that the metal structure after hot forging has a certain directivity, usually called forging streamline, also known as streamlines. The forging streamline makes the metal properties anisotropy, and the tensile strength along the streamline direction (longitudinal) is high, while the tensile strength perpendicular to the streamline direction (transverse) is low. In production, if the streamlined structure in the forging can be continuously distributed and consistent with its stress direction by taking advantage of its high longitudinal strength, the bearing capacity of the part will be significantly improved. For example, when the hook is formed by the bending process, the streamlined direction can be consistent with the force-bearing direction of the hook, as shown in Figure 4-17a), which can improve the ability of the hook to withstand the tensile load. The streamlined distribution of the forged crankshaft shown in 4-17b) is reasonable. Figure 4-17c) shows the crankshaft formed by cutting. Because the streamline is discontinuous, the streamlined distribution is unreasonable.

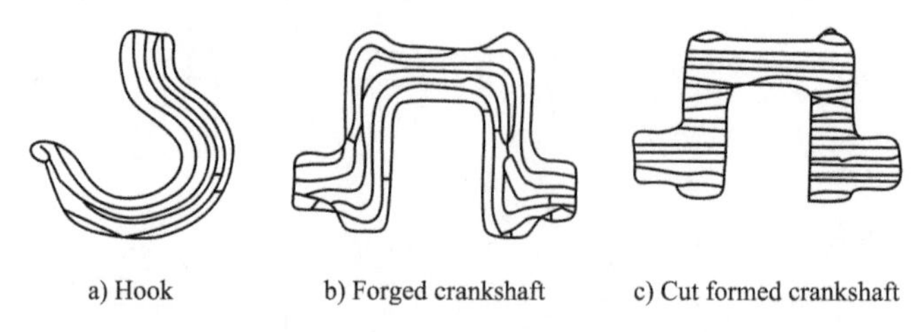

a) Hook b) Forged crankshaft c) Cut formed crankshaft

Figure 4-17 Streamline distribution in hook and crankshaft

(2) Forging ratio.

The forging ratio is a parameter indicating the degree of metal deformation. The forging ratio is closely related to the forging process. When it comes to the drawing process, the forging ratio is represented by $Y_{drawing}$, and when it comes to the upsetting process, the forging ratio is represented by $Y_{upsetting}$. The specific calculation formula is as follows:

$$Y_{drawing} = S_0/S \quad (4\text{-}1)$$

$$Y_{upsetting} = H_0/H \quad (4\text{-}2)$$

Where, S_0——the cross-sectional area of the metal blank before drawing;

S——the cross-sectional area of the metal blank after drawing;

H_0——indicates the height of the metal blank before upsetting;

H——the height of the metal blank after upsetting.

4.3.2 Common Pressure Processing Methods

According to different sources of force during forging, forging can be divided into manual forging and machine forging. Manual forging is forging tools by hand, operating on iron and stone. This simple and crude production method is only used for repairing work or small batch production. Machine forging is a forging method that relies on various forging equipment to provide force, and it is the main style of modern forging production. Machine forging can be divided into free forging, mold forging, hammer die forging, etc.

1) Free forging

Forgings that only use simple tools or directly deform the blank between the upper and lower anvils of the forging equipment to obtain the required geometric shape and internal quality are called free forging.

The technological process for free forging on other forging equipment is composed of a series of basic processes.

(1) Basic process of free forging.

The common processes of free forging can be divided into swaging, upsetting, punching, bending, staggering, and torsion.

① Swaging is a forging process that reduces cross-sectional area and increases the length of the blank. It is commonly used for forging shafts or forgings with long axis lines.

② Upsetting is a forging process that reduces the blank height and increases the cross-section area, and is commonly used for forging gear blanks and round cake forgings.

③ Punching is a forging process in which a punch is used to punch through or through the holes on the blank after upsetting, and is often used to forge hollow forgings such as rods, gear blanks, rings, and sleeves.

④ Bending is a forging process in which a certain tool and die are used to bend the blank into the specified shape. It is commonly used to forge forgings with bent axes such as angle ruler, bending plate, and hook.

⑤ Staggering refers to the forging process in which the T-shape of the blank is staggered in parallel with another part for a certain distance. It is commonly used for forging crankshaft parts. In case of stagger shifting, the billet shall be partially cut first. Then, the impact force or pressure with equal size, opposite direction, and perpendicular to the axis shall be applied on both sides of the notch to realize stagger shifting of the billet.

⑥ Torsion is a forging process that rotates one part of the blank relative to the other part around its axis at a fixed angle. It is commonly used for forging multi turn bending, fried dough twist drill, and correcting some forgings. When the torsion angle of the small blank is small, the hammering method can be used.

(2) Production characteristics and application of free forging.

During free forging, only the blank part contacts with the upper and lower anvils to produce plastic deformation, and the rest is a free surface, so the tonnage of forging equipment is required to be relatively small. The process flexibility of free forging is large. When changing the forging variety, the production preparation time is short, the productivity of free forging is low, the forging precision is not high, and the forging with complex shapes cannot be forged. Free forging is mainly used under single-piece and small-batch production conditions.

2) Mold forging

Mold forging is a forging method that uses a movable mold to produce die forgings on free forging equipment. The tire mold is not fixed on the hammer or anvil, but only placed when it is in use. In the production of medium and small forgings, free forging and mold forging are widely used.

The mold forging process is flexible and there are many kinds of mould. Understanding the structure and forming characteristics of the mould is the key to mastering the mold forging process.

According to the structural characteristics of the mould, the mould can be divided into four types, namely, tumbler, buckle, sleeve and clamping.

Compared with free forging, mold forging has the following advantages.

(1) Because the blank is formed in the mould chamber, the forging size is relatively

accurate, the surface is relatively smooth, and the distribution of streamlined structure is relatively reasonable, so the quality is high.

(2) As the forging shape is controlled by the die web, the blank forming is fast and the productivity is 1 ~ 5 times higher than that of free forging.

(3) The mold forging can forge the forgings with complex shapes.

(4) There are just a few residual forgings, so the machining allowance is small, which can not only save metal materials, but also reduce machining hours.

Mold forging also has some disadvantages: It requires a forging hammer with a large tonnage. Only small forgings can be produced. The service life of the mould is low. When working, the mould must be moved manually, so the labor intensity is high.

3) Hammer die forging

Die forging on the hammer is called die forging for short. It is a forging method that uses a die (forging die) on the die forging hammer to deform the blank and obtain the forging. The tool that forms the blank to obtain a die forging is called a forging die. The forging die can be divided into single-die chamber forging die and multi-die chamber forging die.

Compared with free forging and mold forging, hammer die forging has the following advantages.

(1) High productivity.

(2) The surface quality is high, the machining allowance is small, there are few or no remaining pieces, the size is accurate, and the forging tolerance is 2/3 ~ 3/4 smaller than that of free forging, which can save a lot of metal materials and machining time.

(3) Simple operation, lower labor intensity than free forging and mould forging.

Main disadvantages of die forging on hammer:

① The weight of die forging is limited by the capacity of general die forging equipment, mostly below 50 ~ 70kg.

② The forging die needs expensive die steel, and the processing of the die chamber is relatively difficult, so the manufacturing cycle of the forging die is long and the cost is high.

③ The investment cost of die forging equipment is higher than that of free forging, and mould forging is used to produce large quantities of forgings.

4.3.3 Structural Processability of Forgings

The structural process of forgings refers to the degree of difficulty in forging the designed parts with forgings as blanks, on the premise of meeting the user needs. Different forging methods have different requirements for the structural technology of parts. The structural and technological requirements of free forging are as follows.

(1) The shape of the forging shall be as simple, symmetrical, and straight as possible, so as to adapt to the equipment characteristics that the upper and lower parts of the forging are flat.

(2) Conical and wedge-shaped surfaces shall be avoided on forgings, as shown in Figure 4-18.

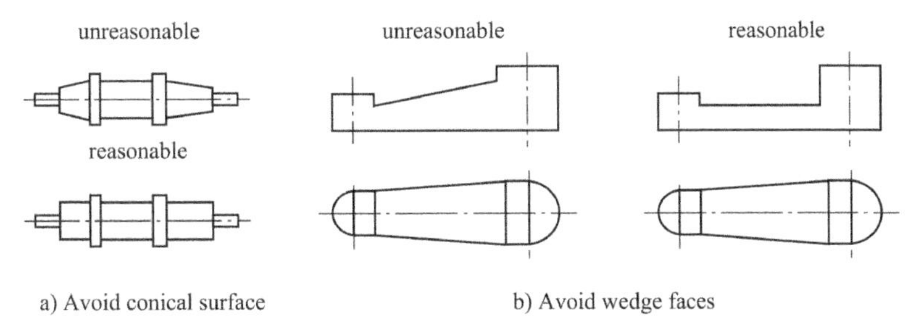

Figure 4-18 Forging structure to avoid conical and wedge surfaces

(3) Avoid the intersection of cylindrical faces and prismatic faces. Because the intersection of these surfaces is a complex curve, it is difficult to forge, as shown in Figure 4-19.

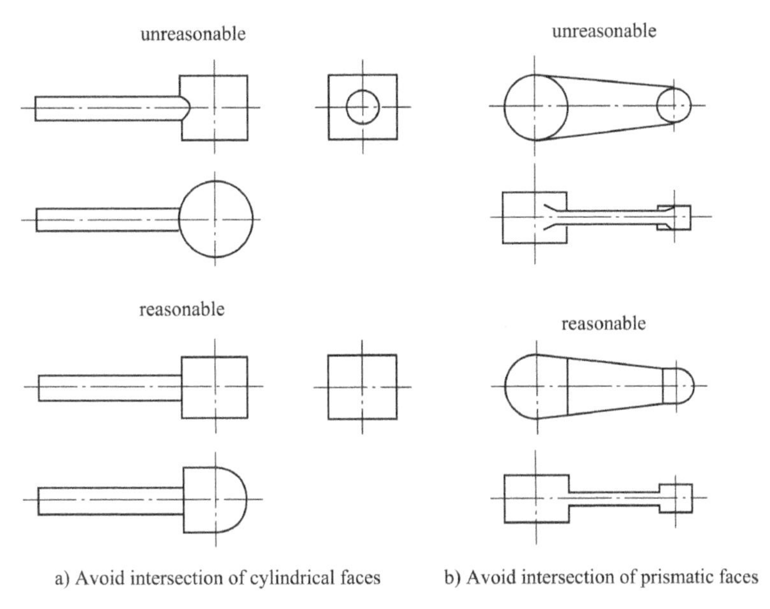

Figure 4-19 Structure to avoid complex curves on forgings

(4) There shall be no reinforcing rib on the forging. It is correct to use reinforcing ribs on castings to improve the bearing capacity of parts, and it will not be too difficult to produce ribbed castings by the casting method. However, it is obviously unreasonable to set the reinforcing rib on the free forging, because it is impossible to forge the rib on the flat anvil. The reasonable way is to increase the diameter or wall thickness of the part, as shown in Figure 4-20.

(5) Avoid bosses on forgings. Because the boss cannot be made by free forging. If the flange plate with four bosses shown in Figure 4-21a) is changed to the fisheye pit structure shown in Figure 4-21b), it will not be too difficult to forge it, because these fisheye pits can be forged after adding surplus pieces.

a) With reinforcing rib　　b) Without reinforcing rib

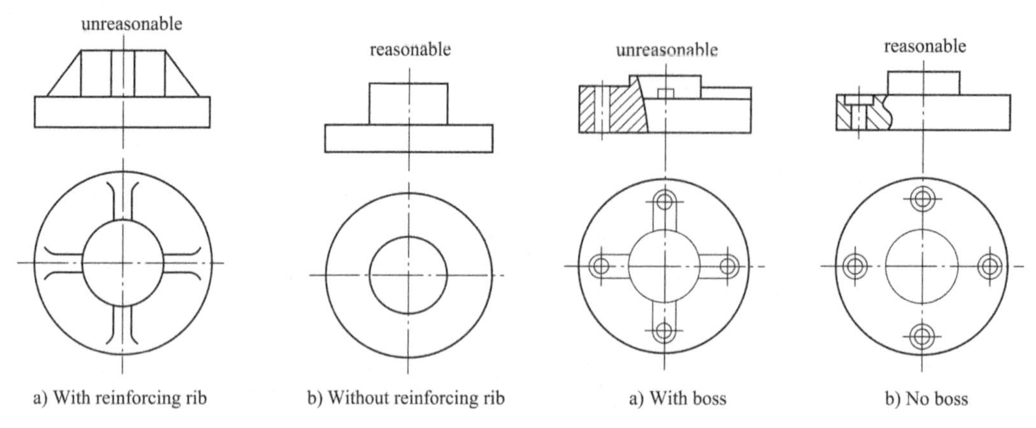

Figure 4-20　Forging structure with or without stiffening ribs

a) With boss　　b) No boss

Figure 4-21　Method of improving the structure of small boss

(6) The assembly structure is adopted. For parts with a large difference in section size and parts with complex shapes, it can be considered to divide the parts into several parts with a simple shape, forge them separately, and then connect them into a whole by welding or threaded connection.

4.4　Welding Processing and Its Application

In modern industrial production, it is often necessary to connect several parts or materials together. Common connection methods include key connection, bolt connection, chat connection, welding, adhesive connection, etc. The first two connection modes are mechanical connections, which can be disassembled. The last three connection modes belong to permanent connections and are non-detachable. Among them, welding is the most widely used connection.

Welding is widely used in the manufacturing of bridges, vessels, ships, boilers, hoisting machinery, television towers, metal trusses, and other structures. With the development of welding technology and the application of computer technology in welding, the welding quality and productivity will also be continuously improved, and the application of welding in national economic construction will be more extensive.

4.4.1　Basic Principles of Welding and Common Welding Methods

1) Classification of welding

Welding refers to a processing method that can make the workpieces achieve non-detachable connection by heating or pressurizing, or both, with or without filler materials. Non-detachable connection refers to the condition that the connected parts must be destroyed or damaged before they can be disassembled. Objects that are welded are called weldments. The products produced by welding are various welding structures. Welding structure refers to the metal structure obtained by connecting the weldments by welding.

There are many welding methods, which can be divided into three categories according to their process characteristics.

(1) Fusion welding. In the welding process, the welding method is to heat the weldment joint to the molten state, and then connect the separated workpieces into a whole after cooling and crystallization.

(2) Pressure welding. In the welding process, the weldment must be applied with pressure (heating or not) to complete the welding.

(3) Brazing. The filler metal with a lower melting point than the weldment is used to heat the weldment together, while the weldment does not melt. After melting, the filler metal is filled into the gap between the weldment and the weldment to be soldered.

After solidification, two weldments are connected into a whole. Brazing can use flame or electricity as the heating source. To improve the wettability and remove the oxide film, solder should be used. According to the different melting points of the solder used, it can be divided into brazing (the melting point of the solder is greater than 450℃) and soldering (the melting point of the solder is less than 450℃).

2) Welding characteristics

Welding production has the following characteristics.

(1) Reduce the structural mass and save metal materials. Compared with the traditional connection method - riveting, welding can save 15% ~20% of metal materials. The self-weight of the metal structure is also reduced due to the saving of materials.

(2) Bimetallic structure can be manufactured. Welding can be used for butt welding and friction welding of different materials. It can also be used to manufacture composite vessels to meet the special performance requirements of high-temperature, high-pressure equipment, chemical equipment, etc.

(3) It can transform the big part into the small part, and connect the small part to the big part. When manufacturing structural members with complex shapes, the materials can be decomposed into smaller parts first, and then the small parts can be assembled and welded gradually. For large structures (such as the manufacturing of ship hulls), small ones are used to make large ones.

(4) High structural strength and good product quality. In most cases, the welded joint can reach the same strength as the base metal, even the joint strength is higher than the base metal strength. Therefore, the product quality of the welded structure is better than that of riveting. At present, welding has mostly replaced riveting.

(5) The noise during welding is low, and the labor intensity of workers is low. High productivity, easy to realize mechanization and automation.

(6) It will produce welding force and deformation. Since welding is an uneven heating and cooling process, welding stress and deformation will occur after welding. If certain measures are

taken during welding, the welding force and deformation can be eliminated or reduced.

4.4.2 Process Basis of Fusion Welding

In the welding process, the method of heating the weldment joint to the molten state and completing the welding without pressure is called fusion welding. Common fusion welding methods include arc welding, electroslag welding, and gas welding. Among them arc welding is the most widely used. Arc welding can also be divided into shielded metal arc welding, submerged arc welding, and gas shielded welding.

1) Shielded metal arc welding

Shielded metal arc welding (SMAW), also known as manual metal arc welding (MMA or MMAW), flux shielded arc welding or informally as stick welding, is a manual arc welding process that uses a consumable electrode covered with a flux to lay the weld.

(1) Welding electrode.

The welding arc is supplied by the welding power source. It is a strong and lasting discharge phenomenon in the gas medium between two electrodes with a certain voltage or between the electrode and the weldment.

① Basic structure and heat distribution of welding arc.

The structure and heat distribution of the welding arc can be divided into three areas according to the welding arc, as shown in Figure 4-22, namely the cathode area, anode area, and column grabbing area. When the DC power supply is used, if the electrode is connected to the negative electrode and the working speed is set to the positive electrode, the cathode area is at the end of the electrode and the anode area is on the workpiece.

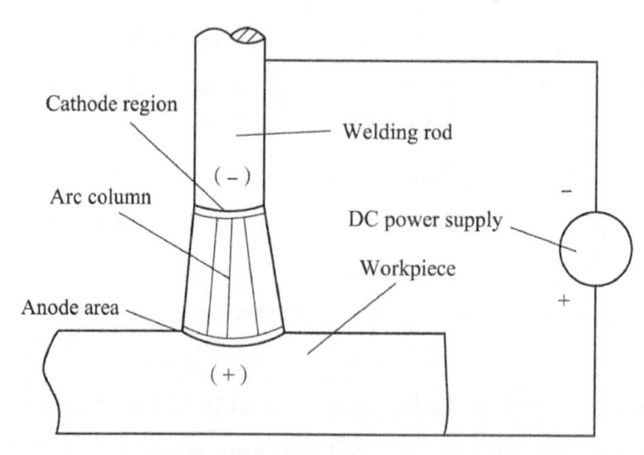

Figure 4-22 Composition of the welding arc

Cathode area refers to the narrow area near the cathode end, which is the origin of primary electron emission. Anode area refers to the area near the anode end, which is the area where the anode surface receives electrons during arc discharge. The gas space area between the cathode area and the anode area is the arc column area, and its length is equivalent to the length of the entire arc. When welding steel with steel electrodes, the heat released from the cathode area accounts for about 36% of the total heat of the arc, and the temperature is about 2100℃. The heat released from the anode area accounts for about 43% of the total heat of the arc, and the temperature is about 2300℃. The heat released from the arc column area accounts for about 21% of the total heat of the arc, and the central temperature of the arc column can reach above 5700℃. When using an AC welding power source, the temperature of the two poles is basically the same because the polarity of the power source changes rapidly and alternately.

② Welding current electrode.

When the DC power supply is used for welding, the workpiece is connected to the positive pole of the power supply and the welding rod is connected to the negative pole, which is called positive connection. If the workpiece is connected to the negative electrode, the electrode is connected to the positive electrode, called reverse connection. When welding thin parts, if the positive connection method is adopted, the welding parts will have burn-through defects due to large heat and high temperature. When welding thick plates, if the reverse connection method is adopted, the weldment will have the defect of incomplete penetration due to low heat and low temperature. Therefore, when the DC welding power supply is used, the positive and negative electrodes should be connected according to the thickness of the weldment.

In general, the reverse connection shall be adopted when welding thinner weldments. If thicker parts are welded, the positive connection shall be adopted. When welding with the AC power supply, there is no positive or reverse connection problem.

(2) Welding rod.

Welding rod is an important welding material in electrode electric grasping, which is composed of a welding core and coating. It directly affects the stability of the welding arc, chemical composition, and mechanical properties of weld metal. The quality of the welding rod is one of the main factors affecting the quality of arc welding with the welding rod.

① Welding core.

The electrode core is referred to as the welding core, which is the metal core covered by the coating in the electrode. During welding, the welding core has two functions: one is to conduct current and generate electric arc. The second is that the welding core is melted during welding, which is used as the filler metal to fuse with the molten liquid metal of the weldment to form a weld. The core metal accounts for about 50% ~ 70%. It can be seen that the chemical composition of the core has a great impact on the chemical composition of the weld metal.

② Drug skin.

In the process of shielded metal arc welding, if the so-called bare electrode welding is carried out directly with the welding core, a large amount of oxygen and nitrogen will invade the molten metal during the welding process, oxidize and nitride the metal iron and beneficial elements such as carbon, silicon, pickaxe into various oxides (such as FeO) and nitrides (such as Fe4N), and remain in the welding key, causing slag inclusion defects in the weld. The gas dissolved into the flame pool can cause a large number of pores in the weld. In this way, the mechanical properties (strength, impact toughness, etc.) of the weld are greatly reduced. In addition, due to the lack of easily ionized materials, the welding arc is unstable, the spatter is serious, and the weld formation is poor.

In order to prevent the above defects, a layer of coating can be coated on the outside of the welding core. In the welding process, the coating plays a complex metallurgical reaction and physical and chemical changes, basically overcoming the problems that occur in light electrode welding. Therefore, the coating is also one of the main factors affecting the weld metal quality.

The main functions of the coating are as follows: First, mechanical protection. During welding, the coating and the welding core melt together, generating a large amount of reducing gas and low melting point slag through a series of physical and chemical reactions. Both of them can play a mechanical isolation role to prevent harmful gases from invading the molten metal of the weld. Second, metallurgical treatment. Through the metallurgical reaction between the slag and the molten metal, harmful impurities (such as oxygen, hydrogen sulfur, and phosphorus) can be removed. The beneficial alloy elements are added to make the weld obtain the required mechanical properties. The third is to improve the welding process performance. Because the coating contains substances that are easy to ionize, it can make the arc stable, spatter less, easy to operate, weld formation good, and get a flat and dense weld.

2) Submerged arc welding

The welding method of arc burning under the flux layer is called submerged arc mound, which is divided into buried grasp welding and manual submerged arc welding, among which manual submerged arc welding is widely used in production.

(1) The forming process of submerged arc welding seam.

Figure 4-23 shows the principle of the submerged arc welding process. Before welding, cover the welding head with a layer of granular flux with a thickness of 30 ~ 50 mm, and then insert the welding wire into the flux to keep a proper distance from the joint of the weldment and make it generate an arc. The heat generated by the electric grip melts the surrounding flux into slag and forms high-temperature gas. The slag on both sides of the high-temperature gas is arranged to form a cavity, and the arc burns in this cavity. The arc is isolated from the outside air by the liquid slag covered on it and the most unmelted flux on the surface. After the welding wire melts, it forms a molten drip, which is mixed with the melted weldment metal to form a molten pool. As the welding wire moves in the direction indicated by the arrow, the liquid metal in the molten pool solidifies and forms a weld. At the same time, the slag floating in the molten pool also solidifies into a slag shell.

(2) Characteristics and applications of submerged arc welding.

Compared with electrode arc welding, submerged arc welding has the following advantages:

① Good welding quality. Because the welding process can be controlled automatically, the process parameters can be adjusted to the optimal value, the chemical composition of the weld is more uniform and stable, and the weld forming is smooth. At the same time, harmful gas is difficult to invade, metal metallurgy reaction of molten pool is sufficient, and welding defects are less.

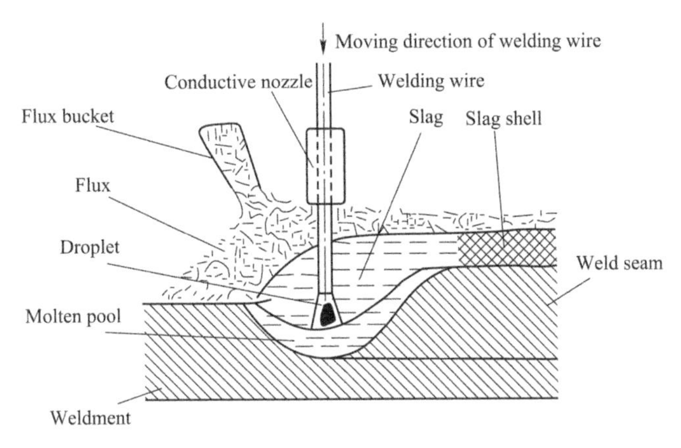

Figure 4-23 Principle of the submerged arc welding process

② High productivity. Due to the short length of the welding wire protruding from the conductive nozzle, a larger welding current can be used. This will increase the welding speed. At the same time, when the thickness of the weldment is less than 14 mm, the groove is not needed, so the welding productivity is high.

③ Save welding materials. Because only small groove or no groove is needed for submerged arc welding, it can reduce the filling amount of welding wire in the weld, and also reduce the consumption of weldment material. At the same time, because of the small metal splash during welding, and no loss of electrode head, so can save welding materials.

④ Easy to realize automation, low labor intensity, better working conditions, simple operation.

The disadvantages of submerged arc welding are as follows: high equipment cost. Due to the use of granular flux, generally can only be welded flat weld, and is not suitable for welding complex structure or inclined weld welding parts. Because the arc can not be seen, it is not convenient to check the weld quality during welding.

Submerged arc welding is suitable for long weld welding of thick plates of low carbon steel, low alloy steel, stainless steel, copper, aluminum, and other metal materials.

3) Gas shielded arc welding

Arc welding with external gas as the arc medium and protect the arc and welding area is called gas shielded arc welding, short for gas shielded welding. The most commonly used gas shielded arc welding methods are argon arc welding and carbon dioxide gas shielded welding.

(1) Argon arc welding.

Argon arc welding is an electric arc welding using argon gas as the shielding gas. Argon arc welding is divided into two types: molten arc welding and non-molten arc welding, according to whether the electrode melts during the welding process, see Figure 4-24a) and Figure 4-24b), respectively. Fused electrode argon arc welding is a welding method that uses a solid core wire with a diameter of ϕ0.8 to 2.44mm and is protected by argon gas to protect the arc and molten pool. The wire is both the electrode and the filler metal, so it is called fusion electrode argon arc welding. Non-melting electrode TIG welding is gas-shielded welding with the tungsten electrode as the electrode and argon gas as the shielding gas. In the welding process, the tungsten electrode does not melt, so it is called non-melting electrode argon arc welding. The filler metal is fed into the arc area by melting the wire. Compared to other arc welding methods, argon arc welding without flux to obtain a high-quality weld. Since it is open arc welding, it is easier to operate and observe and can be welded in various spatial positions.

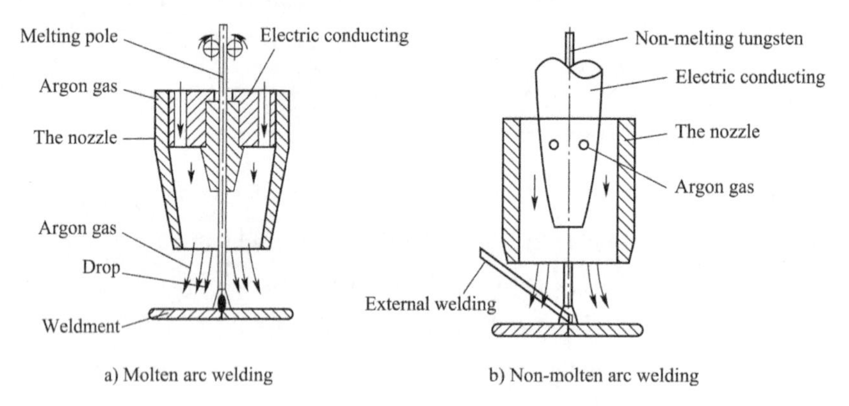

Figure 4-24 Argon arc welding diagram

Argon arc welding is mostly used for welding materials such as aluminum, magnesium, titanium, copper and its alloys, low-alloy steel, stainless steel, and heat-resistant steel.

(2) Carbon dioxide gas welding.

Carbon dioxide gas shielded welding is a fusion arc welding in which carbon dioxide is used as a shielded gas for welding while the solid core photowelding wire is continuously sent out, as shown in Figure 4-25.

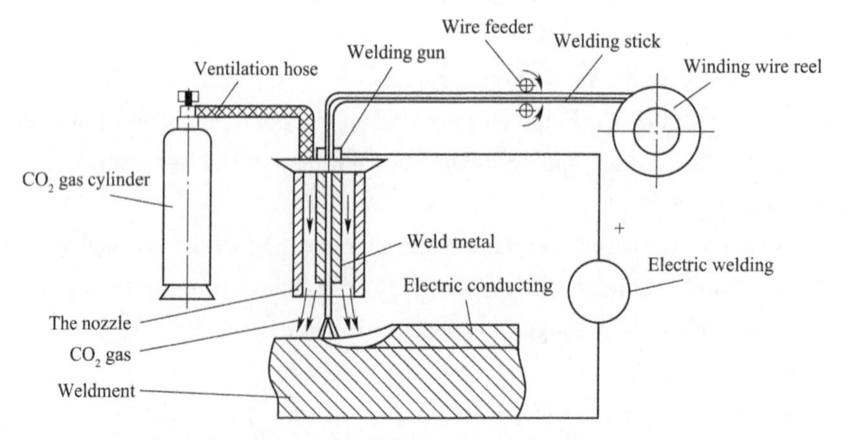

Figure 4-25 Schematic diagram of carbon dioxide gas shielded welding

The advantage of carbon dioxide gas welding is high productivity. Carbon dioxide gas is cheaper than oxygen, and carbon dioxide gas welding consumes less electricity, so the cost is lower. Because of the arc heat concentration, the weld pool is small, the weldment deformation is small, and the welding quality is high. Its disadvantages, are that it is not suitable for welding easily oxidized non-ferrous metals and other materials, nor should it work in windy sites, the electric arc light is strong, the melt drop splash is serious, and weld forming is not smooth.

Carbon dioxide gas welding is commonly used for welding carbon steel, low alloy steel, stainless steel, and heat-resistant steel. It is also suitable for repairing machine parts, such as the surfacing of zero-wear parts.

4.4.3 Weldability of Metal Materials and Welding of Common Metal Materials

In the design of a welding structure, it is necessary to understand what welding materials can be used to obtain high-quality welding quality, or when certain welding materials are known to be difficult to weld, what measures should be adopted to ensure the quality of welding, that is to say, only by understanding the weldability of metal materials, can we correctly carry out welding structure design, preparation before welding, and draft welding process.

1) Weldability of metals

The weldability of metal refers to the adaptability of metal materials to the welding process. It mainly refers to the difficulty of obtaining high-quality welded joints under certain welding process conditions. It includes two aspects: one is the process performance, which means the sensitivity of metal to forming welding defects (mainly cracks) under certain welding process conditions. The other is the performance, that is, under certain welding process conditions, the adaptability of metal welded joints to the requirements of the user.

In low-carbon steel welding, it is easy to obtain a defect-free welded joint without taking complex technological measures. If the same process is used to weld cast iron, it will often produce cracks, which result in low-quality welded joints. Therefore, mild steel's weldability is better than cast iron's.

Complete welded joints do not always have good performance. For example, when welding cast iron, even cracks and other defects are not found, but because of the easy formation of white tissue in the fusion zone and semi-fusion zone, it can not be used because of processing and brittleness. This means that cast iron is not weldable well.

2) Weldability of commonly used metallic materials

(1) Weldability of mild steel.

Low-carbon steel carbon equivalent is low, has good weldability, and generally does not need to take special technological measures to get high-quality welded joints. In addition, low-carbon steel can be welded by almost any welding method.

Low-carbon steel welding generally does not need to be preheated. Preheating is considered only in a cold climate or for thicker weldment. For example, when the sheet thickness is greater than 30mm, or the ambient temperature is lower than −10℃, the weldment needs to be preheated to 100 ~ 150℃.

(2) Weldability of medium-carbon steel.

Medium-carbon steel has a higher carbon equivalent and poorer weldability than low-carbon steel. The heat-affected zone of medium-carbon steel weldments is prone to hardened tissue. When the thickness of the weld is larger, the welding process is not appropriate, making it easy to produce cold cracking. At the same time, the weld joint part of the carbon dissolves into the weld pool so that the weld metal-carbon equivalent increases, reducing the plasticity of the weld, which makes it easy to produce thermal cracking in the solidification cooling process.

The medium-carbon steel should be preheated before welding to reduce the cooling rate of welded joints, the hardening tendency of the heat-affected zone, and prevent cold cracks. The preheating temperature is generally 100 to 200℃.

Medium-carbon steel weldment joints should be beveled to reduce the proportion of weldment metal melting into weld metal and prevent thermal cracking.

(3) Weldability of low alloy structural steel.

Low-alloy structural steel has a greater hardening in the heat-affected zone of the weldment. The strength grade of low-alloy structural steel, low-carbon content, and hardening tendency are small. As the strength grade increases, the carbon content of the steel also increases, coupled with the influence of alloying elements, so that the hardening tendency of the heat-affected zone also increases. Therefore, resulting decreased plasticity of the welded joint, the tendency to produce cold cracking also increased. It can be seen that the weldability of low-alloy structural steel becomes worse with the increase in its strength level.

When welding the low-alloy structural steel, a larger welding current and a smaller welding speed should be selected to reduce the cooling rate of the welded joint. Heat treatment in time after welding or preheating before welding can effectively prevent the generation of cold cracking.

(4) Weldability of cast iron.

The weldability of cast iron is very poor, because its carbon equivalent is large, and the structure has the equivalent of the crack action of graphite. When welding cast iron, the following problems are generally prone to occur.

① It is easy to produce white tissue after welding. Although there are more graphitized elements carbon, and silicon in cast iron, they will be burned seriously in the welding process due to the arc's high-temperature action and the immersion of gas. The reduction of carbon and silicon content, combined with the faster cooling rate, makes the weld prone to the formation of white tissue.

In order to prevent the occurrence of white tissue, the weldment can be preheated to 400 ~

700℃, or the weldment can be insulated and cooled after welding to slow down the cooling speed of the weld. It can also increase the content of graphite elements in the weld metal or non-cast iron welding materials (nickel, nickel-copper, high vanadium steel electrode).

② The crack will occur. Due to the poor plasticity of cast iron, the tensile strength is low. When greater welding stress occurs caused by local heating and cooling in the weldment, it is easy to crack.

Attention should be paid to preheating and slow cooling before and after welding, respectively, to prevent cracks. In addition, the selection of a good plastic electrode (nickel, nickel-copper, high vanadium steel, etc.), eliminating the stress by hammer weld, using a fine electrode and small current, reducing the temperature difference between the weld and the base metal by intermittent welding, etc. can prevent the generation of cracks.

In production, cast iron can not be welded as a material. The welding repair method is used only when there are no serious defects, such as pores, shrinkage holes, sand holes, and cracks on the surface of the cast iron.

4.4.4 Characteristics of Welding Structure and Structure Technology

1) Microstructure and properties of welds

The joint part of the weldment formed after welding is called the weld.

The weld is formed by cooling crystallization of liquid molten pool metal. The crystallization of molten pool metal generally starts from the fusion line at the liquid-solid interface. Since the crystal growth direction is opposite to heat dissipation, the crystal grows from the fusion line to both sides and the center of the molten pool. However, since the growth to both sides is blocked by the adjacent crystals, the crystals mainly grow towards the center of the molten pool. This makes the weld metal obtain a columnar grain structure. Because the cooling rate of the molten pool is faster, the columnar grains are not thick. Coupled with the infiltration of alloying elements in the electrode, the mechanical properties of the weld metal do not change much compared with the base metal.

During the welding process, the microstructure and mechanical properties of the material change due to the influence of heat (but not completely melted) is called the heat-affected zone. The heat-affected zone can be divided into fusion zone, overheated zone, active zone, and incomplete active zone.

2) Welding deformation and prevention methods

(1) Causes of welding deformation.

The internal stress caused by welding is called welding stress, and the deformation caused by welding is called welding deformation. The root cause of welding stress and deformation is the uneven heating and cooling of parts during welding. The primary forms of welding deformation include bending deformation, angular deformation, wave deformation, and twisting deformation, as shown in Figure 4-26.

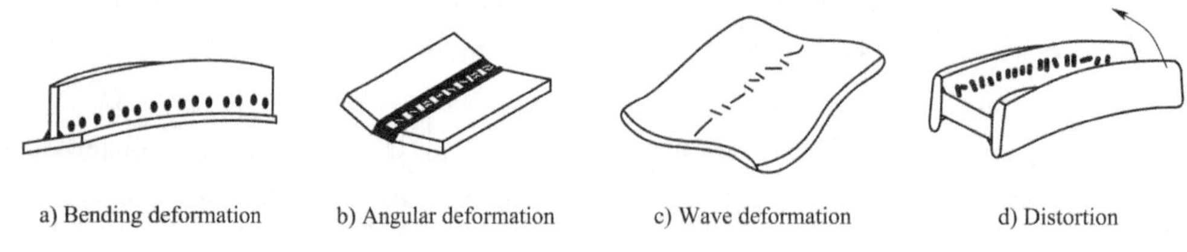

a) Bending deformation b) Angular deformation c) Wave deformation d) Distortion

Figure 4-26 Welding deformation classification

(2) Methods for preventing welding deformation.

① Reverse deformation method. Before welding, layout the weldments in the oppsite direction of their regular deformation pattern to offset the post-weld deformation. Figure 4-27 shows that sheet metal welding makes it easy to produce angular deformation. Before welding, two sheets are placed on the cushion block and bent downward at an angle, which is the angle of upward bending after V-groove welding, so the two sheets are straight after welding.

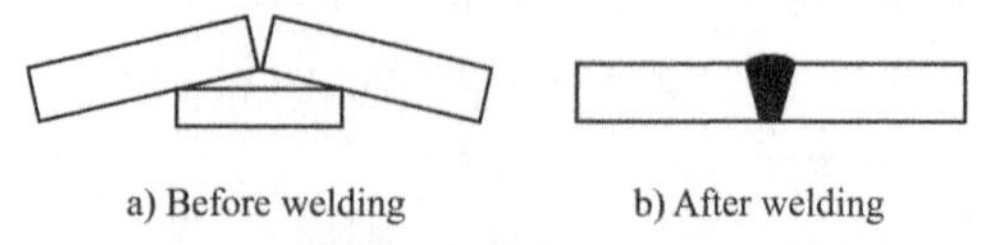

a) Before welding b) After welding

Figure 4-27 Anti-deformation method to prevent corner deformation

② Fixing method before welding. Before welding, pressing plate or weight is pressed on the weldment to resist welding stress and prevent deformation of the weldment, as shown in Figure 4-28a) and b). The welds can also be spot-welded to the platform before welding, as shown in Figure 4-28c). In order to prevent the deformation of the fixed device after removal, it is generally used to knock the weld with a hand hammer during welding to release the welding stress in time, which can make the shape of the weldment more stable.

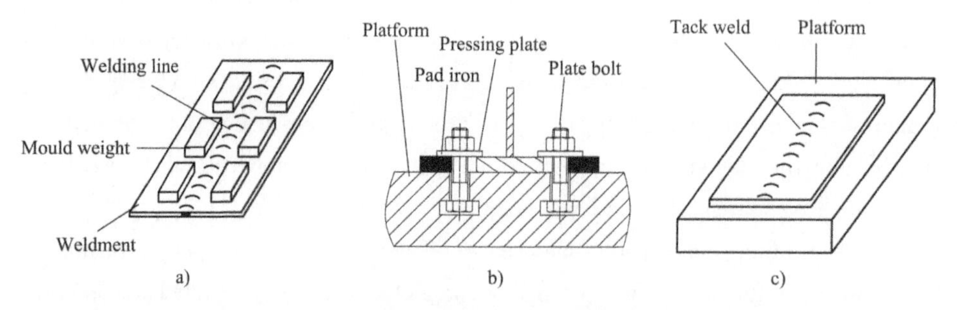

a) b) c)

Figure 4-28 Pre-welding fixation method to prevent deformation

③ Welding sequence changing method. This is a method to prevent welding deformation by changing the order of welding to dissipate the heat applied to the weldment as soon as possible. Common welding sequence transformation methods include the symmetrical welding method, skip welding method, and piecewise backward welding method, as shown in Figure 4-29. The small arrows in the figure indicate the direction of the electrode during welding, and the numbers from small to large indicate the welding sequence.

④ Hammer weld method. This method is a welding method in which a hammer or an air hammer is used to knock the weld metal in order to make the weld metal produce plastic deformation to reduce the welding stress. The striking force should be uniform, and it is best to strike when the weld metal has high plasticity.

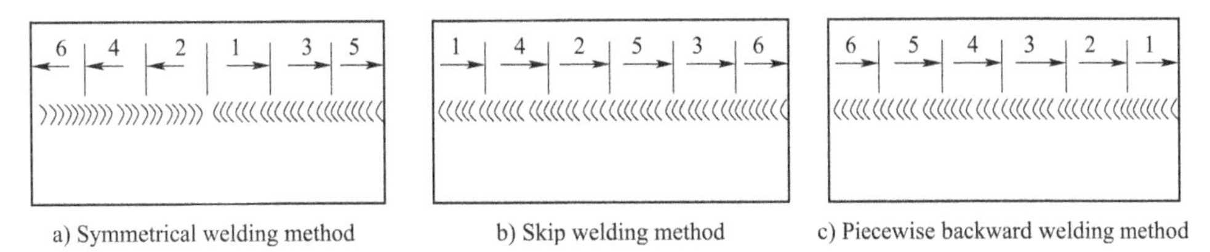

a) Symmetrical welding method b) Skip welding method c) Piecewise backward welding method

Figure 4-29 Welding sequence change method

In the actual production, according to different welding structures, there are many ways to prevent the deformation of the weldment. The above methods to prevent deformation are commonly used, and sometimes several methods can be combined to achieve the best deformation prevention effect.

3) The structure of the weldment technology

Welding structure technology refers to the design of the welding structure on the premise of meeting the performance requirements and the structural welding process requirements, and striving to achieve convenient manufacturing, high productivity, low cost, and good welding quality. Welding technology mainly includes the following three aspects.

(1) Selection of welding structural materials.

The weldability of different metal materials is different, and the corresponding welding process is also different, so the difficulty of welding is different. Therefore, metal materials with good weldability should be selected as far as possible to fabricate weld parts. Generally, low-carbon steel and low-alloy structural steel with low strength grades and good weldability are preferred. Alloys with carbon mass fraction greater than 0.5% and carbon equivalent greater than 0.4% have poor welding performance and are generally unsuitable. Welding structural parts should choose I-beam, channel steel, and various profiles as possible to reduce the number of welding seams, simplify the welding process, and improve the strength and stiffness of the structure. It is difficult to guarantee the welding quality for the welding of dissimilar metal materials due to the different welding performances. The same metal materials should be selected as possible. If necessary, appropriate technological measures should be applied during the welding process. The composed steel has a dense structure and is often used as the material of important welding structures. The boiling steel is easy to produce cracks during welding and is often used as the material of general welding structures.

(2) General principles for weld placement.

In welding structure, weld layout is closely related to welding quality, productivity, cost, and workers' working conditions. Therefore, the following general principles should be followed when considering weld layout:

① The weld arrangement should be dispersed as far as possible to avoid density and crossover and prevent the metal from serious overheating and mechanical property decline, as shown in Figure 4-30.

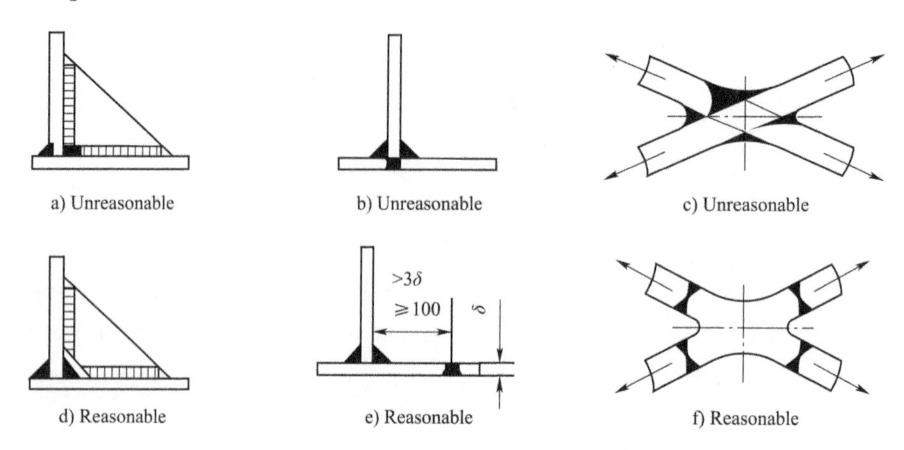

Figure 4-30 Scattered arrangement of welds

② The weld position should be as uniform and symmetrical as possible to reduce welding stress and deformation, as shown in Figure 4-31.

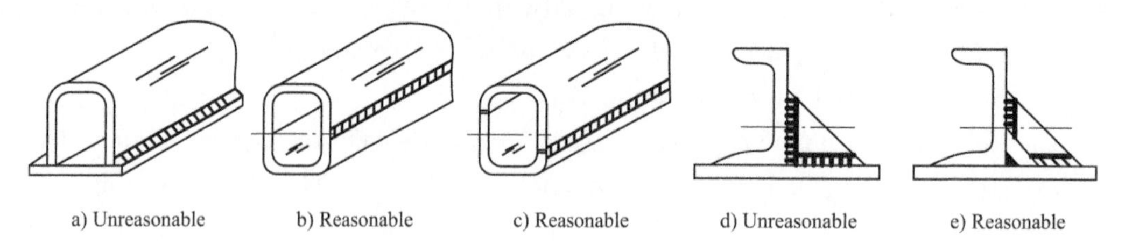

Figure 4-31 Symmetric layout of weld joints

③ Welding seam should try to avoid the maximum stress and stress concentration parts, as shown in Figure 4-32.

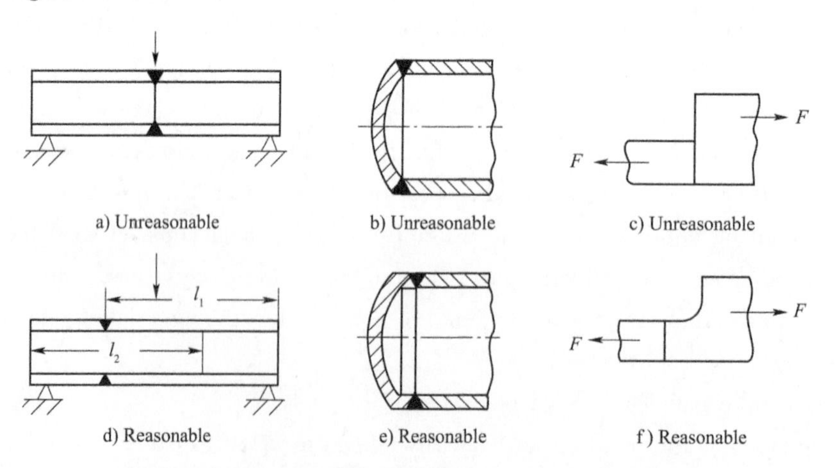

Figure 4-32 Welds avoid the maximum stress and stress concentration

Welding seam should try to avoid the machined surface, which can not affect the accuracy of the machined surface and surface quality, as shown in Figure 4-33.

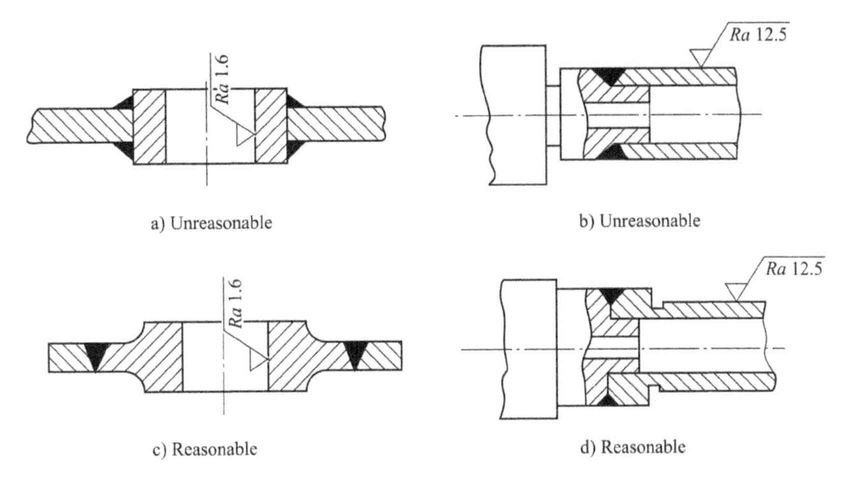

Figure 4-33 The weld is far away from the finished surface

The welding seam should be easy to operate. The operating space of welding should be considered for the electrode arc welding, as shown in Figure 4-34 and Figure 4-35.

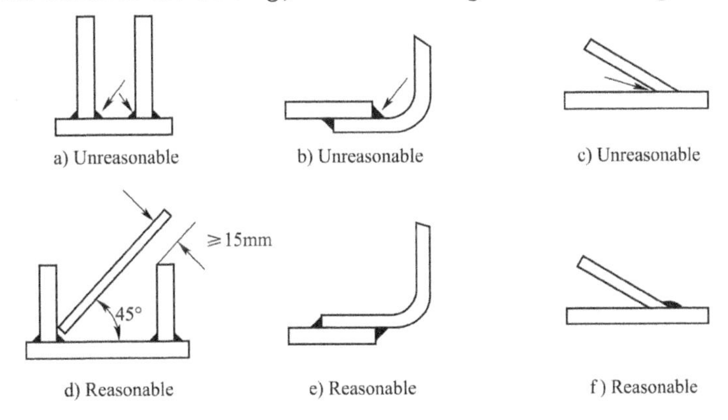

Figure 4-34 Welding position of electrode arc welding

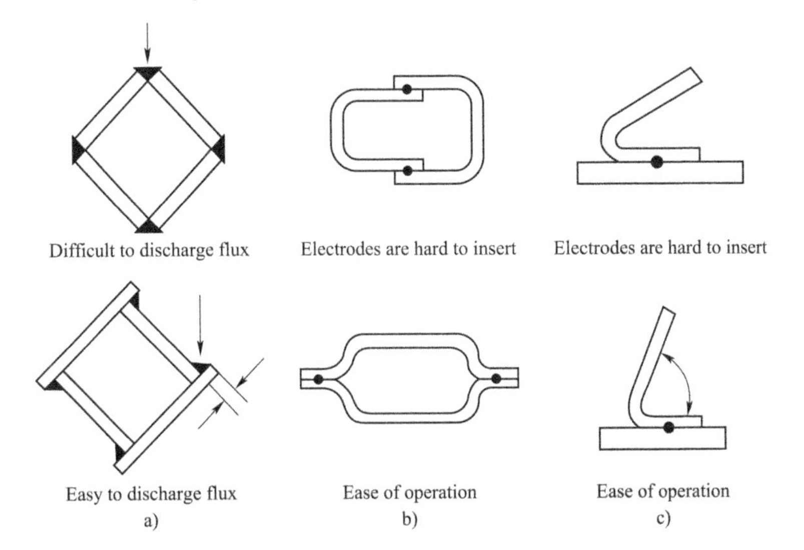

Figure 4-35 Spot welding or seam welding position

The welding bevel design should be reasonable, as shown in Figure 4-36.

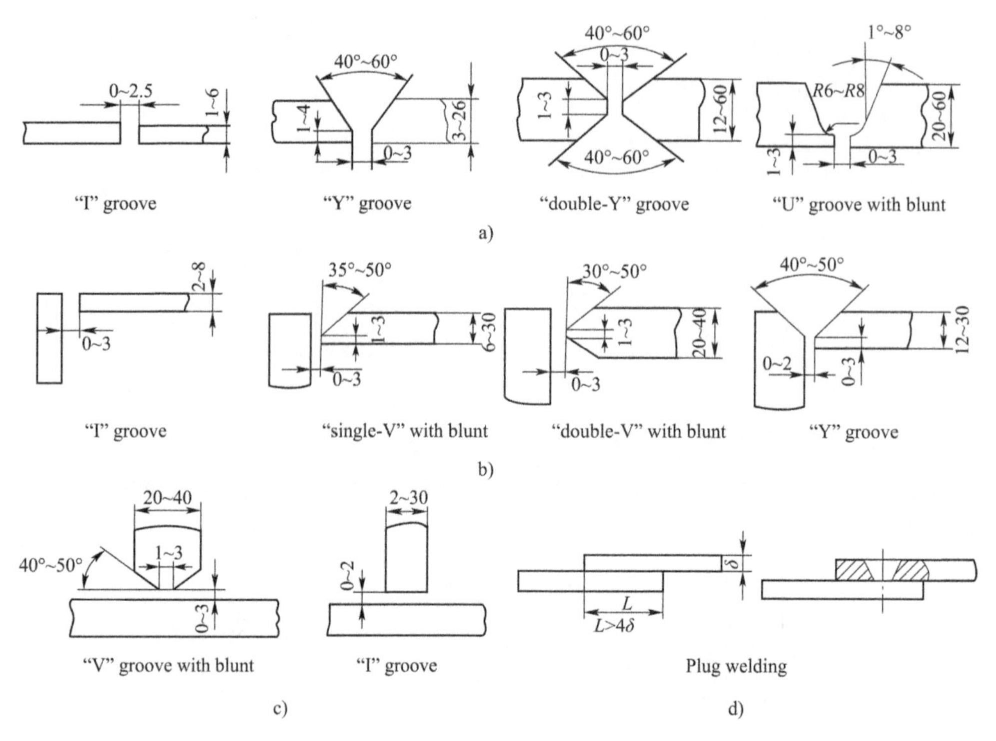

Figure 4-36 Welding groove design

4.5 Conventional Machining Methods and Their Applications

Machining removes unwanted material from a part to produce a specific shape or surface. Conventional machining methods can be divided into cutting and grinding. Cutting is the machining method of removing excess metal materials from metal blanks using cutting tools to obtain mechanical parts meeting the specified technical requirements. A cutting tool used for machining has one or more sharp cutting edges and is made of a harder material than the work material. The cutting edge serves to separate chips from the parent work material. Grinding is a machining method that removes few amounts of material from flat surfaces and cylindrical shapes. Surface grinders reciprocate the work on a table while feeding it into the grinding wheel. The depth to which the wheel cuts usually falls between 0.00025 and 0.001in.

4.5.1 Cutting

Cutting is divided into turning, milling, planning, pulling, drilling, boring, etc.

1) Turning

Turning is a method of cutting the workpiece with the tool rotation on the lathe. Its feature is that the workpiece rotates, and the tool moves. Turning is mainly used to process various rotary surfaces. It is the most widely used processing method in cutting because the rotary surface is used most in mechanical parts.

The technological characteristics of turning are: easy to ensure the position accuracy of each machined surface; suitable for finishing non-ferrous metal parts; high production efficiency; low production cost.

2) Drilling and boring

The primary methods for machining the inner hole surface are drilling and boring. Generally, smaller holes are drilled, larger holes are bored, and large workpieces or holes with higher positional accuracy are machined by boring.

(1) Drilling.

The cutting process carried out on the drilling machine is called drilling, characterized by the tool rotating to make the main motion and moving along the axis to make the feed motion, while the workpiece does not move. The work that can be carried out on the drilling machine includes drilling, expanding, reaming, dimpling hole, and dimpling end-face, etc.

The drilling process features are as follows.

① When drilling, the drill bit is prone to deviation. Deviation refers to the enlargement of the hole diameter, out of the round hole, or deviation of the hole axis caused by the bending of the drill bit during processing. The drill sleeve can be used to guide the drill bit for drilling and reaming to avoid deviation.

② It is difficult to remove chips during processing, and it is easy to scratch the surface of the machined hole when removing chips, reducing the surface quality of the hole, which belongs to rough machining.

③ The cutting heat is not easy to spread, and the drill bit is easy to wear, which limits the improvement of cutting parameters and productivity.

(2) Boring.

Cutting on a boring machine is called boring. The cutter rotates to make the main cutting movement, and the cutter or workpiece makes the axial feed movement.

Boring can correct the axis deviation of the original hole and other errors to obtain higher hole position accuracy. Hence, it is especially suitable for the hole system processing of box workpieces with high accuracy requirements.

3) Planing, slotting, broaching, and milling

(1) Planing, slotting, and broaching.

Planing, slotting, and broaching are mainly used for processing horizontal surfaces, vertical surfaces, internal or external grooves, and formed surfaces. It is characterized by the relative motion path of the tool and the workpiece being a straight line.

① In planing, the relative reciprocating linear motion of the tool to the workpiece is the main motion, the intermittent motion of the workpiece relative to the tool in the direction perpendicular to the main motion is the feed motion, and the planing is carried out on the shaper or gantry planer. Planing is mainly used for machining plane, vertical plane, inclined plane, straight groove, V-shaped groove, dovetail groove, T-shaped groove, and formed surface.

During planing, the main motion is reciprocating, and the cutting process is discontinuous. Due to the influence of inertial force, the cutting speed cannot be very high, and a considerable part of the time is spent on the empty return without cutting, so the production efficiency is low. However, planing has unique advantages like good adaptability, low process cost, convenient processing of long and narrow plane and thin plate plane, and high processing accuracy.

② The slotting process retains the same principle as the planing process, whish is performed on a slotting machine. The difference is that the slotting tool moves the workpiece in a vertical reciprocating linear main motion, therefore the slotting machine is also referred to as a vertical planer. Slotting is mainly used to process the formed inner and outer surfaces of workpieces, such as square holes, polygonal holes, spline grooves, internal gears and external gears. On account of its low production efficiency and low machining accuracy, it is only suitable for single-piece small-batch production and repair processing. The key slot holes or formed holes produced in large quantities are mostly machined by broaching.

③ Broaching is carried out on the broaching machine, which can be used to process through holes of various cross section shapes, and external surfaces of straight line or curve shapes. Broaching has the advantages of high dimensional accuracy and small surface roughness. However, broaching is only applicable to batch or mass production because broaching is a specialized forming tool with complex structure and high manufacturing cost.

(2) Milling.

In addition to planing, milling is more often used for the processing of planes, grooves and steps.

Milling refers to the use of rotating multi-edge tools to cut the workpiece. When milling, the milling cutter rotates to cut the workpiece, which is the main movement. The workpiece is clamped on the workbench and can move forward and backward, up and down, left and right in a straight line or compound curve. It is a feed movement.

The advantages of milling are: the milling cutter is a multi-blade cutter, there is no empty stroke in processing, and the cutting speed is fast, so the production efficiency is higher than planing. As there are many types of milling cutter and many accessories of milling machine, the range of milling processing is broad, and many formed surface processing that cannot be realized by turning and planing can be performed by milling.

However, the stability of the milling process is poor, which affects the processing quality of the workpiece surface.

4.5.2 Grinding

The process of machining the workpiece surface with the grinding wheel or other abrasive tools is called grinding. Grinding has a wide range of processing, which can process various external circles, internal holes, planes and formed surfaces (threads, gears, splines, etc.), including surface, grinding and honing.

1) Surface grinding

Surface grinding is the surface processing on the surface grinder, which is generally used as the finishing process after planning and milling. There are two ways to grind the surface: grinding around the wheel is called peripheral grinding; Grinding with the end face of the grinding wheel is called end-face grinding, as shown in Figure 4-37.

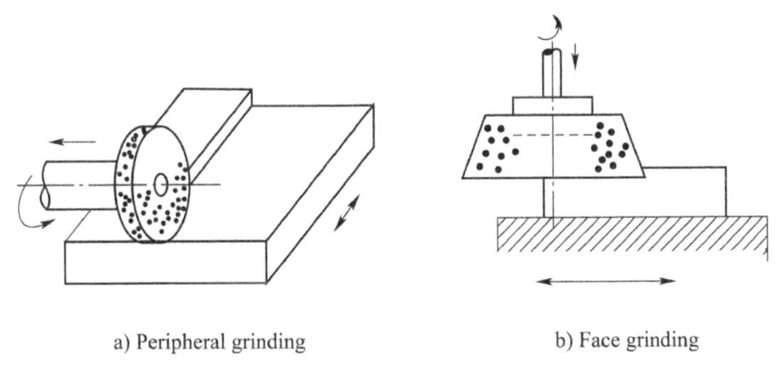

a) Peripheral grinding　　b) Face grinding

Figure 4-37 Surface grinding

(1) With peripheral grinding, the grinding wheel is in line contact with the workpiece, the cutting force is minimal, the grinding efficiency is low, the wear on the grinding wheel's circumference is essentially constant, and the machining precision is high. Face grinding has a high grinding efficieny since the grinding wheel is in face contact with the workpiece; nevertheless, the machining precision suffers because of the uneven wear at the grinding wheel's various points.

(2) Surface grinding is the main method of surface finishing.

2) Honing

Honing is mainly used for hole finishing, Figure 4-38a) shows the schematic diagram of truss grinding. The oilstone grinding strip on the honing head contacts the workpiece hole wall under a certain pressure, and the spindle of the machine tool drives it to rotate and make the axial reciprocating linear motion. In this way, the grinding strip cuts a very thin metal layer from the surface of the workpiece. To avoid the repetition of the abrasive track of the grinding strip, the rotation speed of the honing head and the number of reciprocating stroke per minute must be the prime number to each other. Figure 4-38b) shows the motion track of the grinding strip.

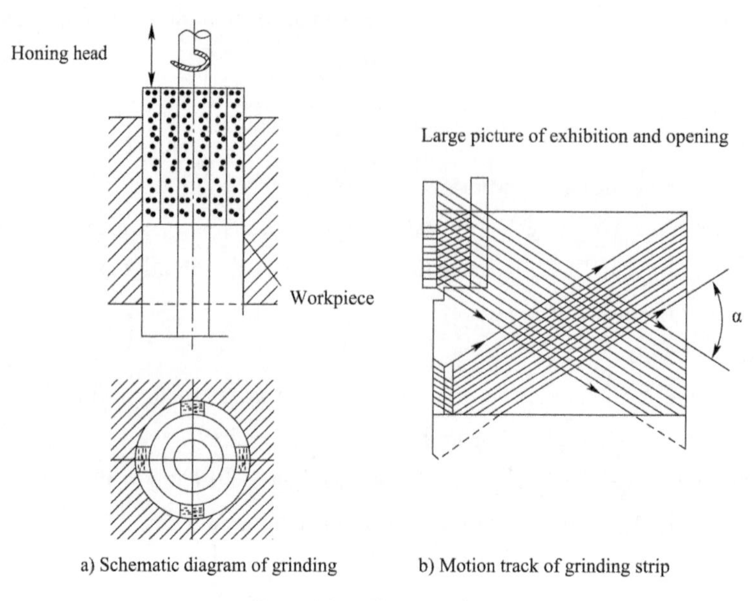

a) Schematic diagram of grinding　　b) Motion track of grinding strip

Figure 4-38　Honing method

Chapter 5 Advanced Manufacturing Technology

5.1 Concept of Advanced Mechanical Manufacturing Technology

Advanced manufacturing technology is an emerging field integrating multiple disciplines of machinery, electronics, control, computer, materials, and management. It is the general term for advanced engineering technology that is based on traditional manufacturing technology, with human as the main body and computer as the main tool, and constantly integrates the latest achievements of machinery, electronics, information, materials, biology and management disciplines, covering all aspects of the entire life cycle of products.

The emergence of advanced manufacturing technology is not only the inevitable result of the development of science and technology, but also the inevitable requirement of civilization and social progress. In the 1970s, the United States emphasized basic research and ignored the development of the manufacturing industry, the manufacturing industry in Japan and Germany developed rapidly. Japan has surpassed the United States in automobile, household appliances, semiconductors and steel, leading to a significant lag in the competitiveness of American products. Japan's manufacturing industry became the first in the world by adopting new manufacturing technologies and management concepts. In the 1980s, to reverse the decline of the manufacturing industry, the United States put forward the concept of advanced manufacturing technology, abbreviated as AMT, to promote the competitiveness of the United States manufacturing industry and the rapid development of the national economy. Subsequently, the major industrial countries in the world, Germany, France, Italy, the United Kingdom and so on, began to study the theory of advanced manufacturing technology. Advanced manufacturing technology, as a high and new technology, is developed based on traditional manufacturing technology by integrating the latest achievements of other disciplines. The purpose is to achieve quality, efficiency, environmental protection, cleanness, low energy consumption, agility and flexibility in the entire manufacturing process. At present, advanced manufacturing technology has become the focus of international scientific and technological competition, and its development level can represent a country's scientific, technological and economic level to a certain extent.

Advanced manufacturing technology has the following characteristics:

(1) Progressiveness. Advanced manufacturing technology is developed on the basis of continuous integration of the latest achievements of other disciplines, which are symbols of the development level of the times.

(2) Extensiveness. Advanced manufacturing technology is no longer limited to the field of manufacturing, but covers the whole process of manufacturing.

(3) Practicability. Advanced manufacturing technology is not aimed at pursuing high and new technology, but has clear economic needs. It is oriented to industrial production, market-oriented, and ends up in the best economic interests of enterprises.

(4) Integration. The development of advanced manufacturing technology has integrated other disciplines, and the boundary between it and other disciplines has gradually faded and disappeared, becoming a new interdisciplinary discipline.

Advanced manufacturing technology can be summarized as advanced manufacturing process technology, manufacturing automation technology, and advanced manufacturing mode.

5.2 Advanced Manufacturing Process

Advanced manufacturing process includes high-speed cutting, special cutting, precision cutting, micro-cutting ,etc.

5.2.1 High-Speed Cutting

1) Concept and development

High-speed cutting technology refers to the advanced processing technology using super hard material tools and abrasives, and high-precision, high-automation and high-flexibility manufacturing equipment that can reliably realize high-speed movement to improve the cutting speed and the material removal rate, processing accuracy, and processing quality. Generally, the cutting speed 5 ~ 10 times higher than the conventional one is called high-speed cutting.

In the 1980s, the rapid development of computer-controlled automatic production technology has become a prominent feature of international production engineering. The numerical control rate of machine tools in industrialized countries has reached 70% ~ 80%. With the application of CNC machine tools, machining centers, and flexible manufacturing systems, the speed of the machine tool empty stroke action (such as automatic tool change, loading and unloading, etc.) and the continuity of the part production process are greatly accelerated, also, the auxiliary working hours of mechanical processing are greatly shortened. This makes cutting hours becoming the main part of the total working hours. Therefore, improving the cutting speed and feeding speed is a necessary way to improve productivity. This is the historical background for the rapid development of ultra-high speed machining technology.

Ultra-high speed machining (UHSM) not only improves the production efficiency of the machine tool, but also further improves the processing accuracy and surface quality of the parts, and can solve the processing problems for some special materials which are difficult to deal with by conventional machining methods. Therefore, the advanced processing technology of ultra-high speed machining has attracted the high attention of industry and academia around the worldwide.

2) Characteristics of high-speed cutting technology

(1) Reduce cutting force and tool wear. The high speed of the tool reduces the deformation of the workpiece and the wear of the tool.

(2) Reduce the thermal deformation of the workpiece. During high-speed cutting, a large amount of heat is taken away by the chip, not transferred to the workpiece.

(3) The size and shape accuracy of parts is high. Because of the small step and cutting depth, high-speed machining can obtain high surface quality, and can even eliminate the process of fitter polishing.

(4) The system vibration is small. High-speed machining is used for automobile mold machining at a cutting speed about 10 times higher than that of conventional machining. Because the excitation frequency of the spindle of the high-speed machine tool is far beyond the natural frequency range of the " machine tool workpiece" system, the machining process is stable and impact-free.

(5) Able to process high-hardness materials. According to the high-speed cutting mechanism, the cutting force is greatly reduced during high-speed cutting, and the cutting process becomes easier. High-speed cutting has great advantages in cutting high-strength and high-hardness materials, and can process materials with complex surfaces and relatively high hardness.

3) Application of high-speed cutting

High-speed cutting has been more and more widely used in aerospace, automobile, mold manufacturing, electronic industry, and other fields.

(1) In the field of aerospace, it is mainly to solve the problem of large surplus materials ' removal, processing of thin-walled parts, high-precision parts, and difficult cut materials, and improvement of production efficiency. As shown in Figure 5-1, aluminum alloy thin-walled parts are formed by cutting off 85% material. High-speed cutting increases material cutting efficiency to 100 ~ 180cm^3/min, which is 3times greater than that of traditional method, and can greatly reduce the cutting time.

Figure 5-1 Aluminum alloy thin-walled part

(2) The application of high-speed machining in the field of automobile production is mainly reflected in mold and parts processing. High-speed machining technology can be applied

to a quite wide range of machinable parts, such as servo valves, various pumps, motor shells, motor rotors, cylinder bodies, and molds, etc. It is gradually adopted in the manufacture of casting moulds for automobile parts and injection moulds for interior parts. The agile flexible automatic production line based on high-speed cutting technology is adopted by more and more domestic and foreign automobile manufacturers. The automatic production lines of FAW-Volkswagen Jetta cars and Shanghai Volkswagen Santana cars imported from Germany have applied a lot of modern high-speed cutting technology. Figure 5-2 shows the car mold processed by high-speed cutting technology.

Figure 5-2 Car door and car mold

(3) In the field of mould and tool industry, high-speed cutting can directly cut hardened material moulds, which saves several traditional processes from machining to electric machining, and also saves man-hours. At present, high-speed cutting can achieve very high surface quality (Ra ≤ 0.4pm), which eliminates the process of surface grinding and polishing after electric machining. In addition, the compressive stress state of the machined surface formed by cutting will also improve the wear resistance of the die workpiece surface. In this way, the forging die and casting die can be processed only by high-speed milling. Complex surface machining, high-speed rough machining, and high-speed finish machining after hardening have great development prospects; there is a trend to replace EDM and polishing machining.

5.2.2 Ultra-Precision Machining

1) Concept and development

Ultra-precision machining technology is a new machining technology developed to meet the needs of modern high-tech development. It integrates the new achievements of mechanical technology, measurement technology, modern electronic technology, and computer technology. Ultra-precision machining has two meanings: One is to break through the difficulty of the traditional machining methods, that is, high-precision machining; the second is realizing the boundary of micro-dimension, namely micro-machining processing. It is known that the machining accuracy is higher than 0.1pm and the surface roughness is less than 0.01pm belongs to ultra-precision machining.

Development trend of ultra-precision machining technology.

(1) The mechanism of new ultra-precision machining methods.

The ultra-precision machining mechanism involves the microcosmic world and the internal structure of matter. The available energy sources can be very extensive, including mechanical energy, optical energy, electrical energy, acoustic energy, magnetic energy, chemical energy, nuclear energy, etc. Also, separation processing, combination processing, deformation processing, and growth accumulation processing can be used. Both individual processing methods and complex processing methods such as precision electrolytic grinding, precision ultrasonic turning, precision ultrasonic grinding, mechanochemical polishing, etc. can be adopted.

(2) Develop in the direction of high accuracy and efficiency.

With the continuous progress of science and technology, the requirements for the processing accuracy, efficiency, and quality of products are becoming higher and higher. The ultra-precision processing technology pushes the limit of processing accuracy, which is infinite. The current goal is to achieve nanometer processing accuracy.

(3) Research and develop integrated processing and measurement technology.

It is urgent to research and develop online machining precision measurement technology, which is significant in ensuring product quality and improving productivity.

(4) Online measurement and error compensation.

Due to the high precision of ultra-precision machining, many factors affecting the machining process are also very complex, and it is also hard to continue improving the accuracy of the machining equipment. Therefore, it is necessary to adopt the online measurement method and computer error compensation to improve the accuracy and ensure the machining quality.

(5) Development of new materials.

New materials shall include new tool materials (cutting, grinding) and processed materials. Because the processed material of ultra-precision machining has a great impact on the processing quality, its chemical composition and mechanical properties are strictly required.

(6) Development toward large-scale and miniaturization.

Due to the development of the aerospace industry, large ultra-precision processing equipment is needed to process large optoelectronic devices, such as mirrors on large astronomical telescopes, while the development of miniaturized ultra-precision processing equipment is mainly to meet the needs of the development of micro-electronic machinery and integrated circuits, such as manufacturing micro-sensors, micro-driven components, etc.

2) Main methods of ultra-precision machining

At present, ultra-precision machining mainly includes ultra-precision turning, ultra-precision grinding, and ultra-precision lapping.

(1) Ultra-precision turning.

Ultra-precision single-point diamond turning technology is one of the most widely used ultra-precision machining methods, which can achieve mirror structure after one machining and obtain the best surface roughness ($Ra < 10nm$). It comprehensively utilizes natural single-crystal diamond tools, ultra-precision lathes (with high positioning accuracy and repetitive accuracy), and C-axis controllable rotation technology to realize the processing of a variety of plastic materials and hard and brittle materials, as shown in Figure 5-3. The ultra-precision single-point diamond turning technology can be used to manufacture all kinds of ultra-precision key components such as plane, aspherical (Figure 5-4), concave/convex aspherical, off-axis aspherical mirrors, aspheric arrays, Fresnel mirrors, micro-groove arrays, and polyhedral prisms. Its surface roughness is up to the order of nanometers, realizing the one-time forming of optical quality surfaces. At the same time, it is also widely used in computer hard disk heads, mobile phone screen microlens arrays, and various microgrooves, as well as high-precision molds (CD discs, aspherical mirrors) for injection molding and production of a large number of cheap parts.

Figure 5-3 Ultra-precision single-point diamond turning

Figure 5-4 Aspheric optical part

(2) Ultra-precision grinding.

Ultra-precision grinding technology is developed based on the general precision grinding. Ultra-precision grinding provides not only mirror-level surface roughness but also ensures accurate geometric shape and size. For this reason, in addition to considering various technological factors, there must be reference components with high accuracy, high stiffness, and high damping characteristics, eliminating the impact of various dynamic errors and adopting high-precision detection and compensation methods. At present, the machining objects by ultra-precision grinding are mainly hard and brittle materials such as glass and ceramics. As a nano-scale grinding process, the machine tool is required with high precision and high rigidity, and brittle materials can be ground ductile.

In addition, the dressing technology of the grinding wheel is also very critical. Although grinding has high efficiency, it is difficult to obtain a mirror surface when grinding glass or ceramics. The main reason is that when the grinding wheel is too fine, the grinding wheel surface is easily blocked by chips. The current ultra-precision grinding technology can produce 0.01μm roundness, 0.1μm dimensional accuracy and *Ra* 0.005μm surface roughness cylindrical parts. Plane ultra-precision grinding can produce 0.03μm/100nm plane. Figure 5-5 shows an ultra-precision grinding part.

Figure 5-5 Ultra-precision grinding parts

(3) Ultra-precision lapping.

Both lapping and polishing are processing methods that use abrasive to make the workpiece and grinding tool pass through relatively complex tracks to obtain high quality and high precision. Ultra-precision lapping includes mechanical lapping, chemical-mechanical lapping, floating lapping, elastic launching lapping, and magnetic lapping. The sphericity produced by ultra-precision lapping can reach 0.025μm, and the surface roughness can reach the *Ra* value of 0.003μm. The mirror surface without the deterioration layer can be processed by elastic emission processing, and the surface roughness can reach 0.5nm. The highest precision ultra-precision lapping can produce a flatness of λ/200. The key conditions of ultra-precision lapping are almost vibration-free lapping motion, precise temperature control, clean environment, and fine and uniform abrasive. In addition, high-precision detection methods are also necessary. Ultra-precision lapping is mainly used to process integrated circuit chips and optical planes with high surface quality and high flatness, as well as sapphire windows.

3) Role of ultra-precision machining technology

Advanced manufacturing technology has become one of the most important technologies for a country's economic development. Many countries attach great importance to the level and development of advanced manufacturing technology and use it to carry out product innovation, expand production and improve international economic competitiveness. At present, the United States, Japan, Germany, emphasizing advanced manufacturing technology innovation, always take the leading position in economic development.

There are two major fields for the technical essence of advanced manufacturing technology: ultra-precision machining technology and manufacturing automation. The former pursues machining accuracy and surface quality, the latter includes the automation of product design, manufacturing and management. It is not only an important means to quickly respond to market demand, improve productivity and labor conditions, but also an effective measure to ensure product quality. There is a close relationship between the above two fields. Many precision and ultra-precision machining rely on automation technology to achieve the expected indicators, while many manufacturing automation depends on precision machining to achieve accuracy and reliability. Both of them play an overall decisive role and are the pillars of advanced manufacturing technology.

(1) Ultra-precision machining , to some extent, represents the national manufacturing industry level. The precision, surface roughness, machining size range, and geometric shape that can be achieved by ultra-precision machining present a country's manufacturing technology level. For example, the size of the blunt radius of the cutting edge is an essential technical parameter of the ultra-precision cutting of the diamond tool. Japan claims it has reached 2nm, while China is still at the submicron level, with one order of magnitude gap. Another example is that the ultra-precision grinding of the diamond micro-powder wheel has been used in production in Japan, which has dramatically improved the manufacturing level and effectively solved the problem of low efficiency of ultra-precision grinding abrasive processing.

(2) Precision machining and ultra-precision machining are the foundation and key of advanced manufacturing technology. At present, a large number of computer-aided manufacturing software have been developed in the field of manufacturing automation, such as computer-aided design (CAD), computer-aided engineering analysis (CAE), computer-aided process planning (CAPP), computer-aided machining (CAM), etc., collectively known as computer-aided engineering (CAX). Another example is designed for assembly (DFA) and design for manufacturing (DFM), which are collectively referred to as design for engineering (DFX). Research on computer integrated manufacturing (CIM) technology, production mode, such as lean production, agile manufacturing, virtual manufacturing, clean production, and green manufacturing are also carried out. These are very important and necessary, representing an important aspect of the current high-tech manufacturing technology.

5.2.3 Rapid Prototype Manufacturing

1) Concept and Development

Rapid prototype manufacturing (RP) is a group of techniques and processes used to quickly create physical prototypes or models of a product or part. The primary purpose of rapid prototype manufacturing is to validate and test a design concept before committing to mass production. It is a critical step in the product development process, enabling designers and engineers to quickly validate their ideas, make design improvements, and ultimately bring high-quality products to market with confidence.

The development of rapid prototype manufacturing has evolved over several decades, driven by advancements in technology, materials, and the growing demand for faster and more efficient product development processes. Rapid prototyping can trace its roots back to the 1960s when engineers and researchers began experimenting with various methods for creating physical models directly from computer-aided design (CAD) data. Early techniques included Stereolithography (SLA), developed in the mid-1980s, which used a laser to solidify liquid resin layer by layer to create 3D objects. The 1990s saw the commercialization of rapid prototyping technologies. Companies like 3D Systems and Stratasys began offering SLA and Fused Deposition Modeling (FDM) systems to a wider range of industries. Additive manufacturing (AM) became more popular in 2000s to describe the broader field that includes rapid prototyping. AM emphasizes the ability to produce functional parts directly from digital models. New AM technologies, such as Direct Metal Laser Sintering (DMLS), allowed for the production of metal parts with complex geometries, expanding the scope of applications. Diverse materials and applications emerged in 2010s. The range of materials available for rapid prototyping and additive manufacturing expanded to include metals, ceramics, composites, and even bio-compatible materials. Industries like healthcare started using 3D printing for custom medical implants and prosthetics. Now, rapid prototyping is becoming an integral part of Industry 4.0, with the integration of sensors, data analytics, and digital twin concepts. Digital twins allow for real-time monitoring and optimization of both the design and manufacturing process. The use of artificial intelligence and machine learning in rapid prototyping is helping to automate design iterations and improve the efficiency of the development process.

2) Principle and process of rapid prototype manufacturing

Unlike traditional machining methods, rapid prototyping manufacturing can quickly produce parts with complex shapes without machining equipment (tools, fixtures, molds). Based on the manufacturing concept of "material stacking layer by layer", the complex three-dimensional processing is decomposed into a simple combination of two-dimensional materials. The physical means of light, heat, and electricity (usually laser) are used to realize the transfer and accumulation of materials.

The basic process of rapid prototyping manufacturing technology is shown in Figure 5-6.

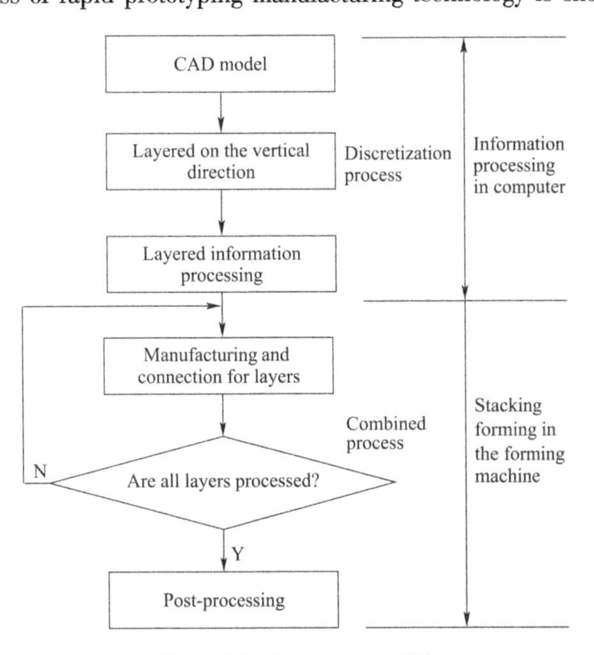

Figure 5-6 Basic process of RP

(1) CAD modeling.

Build 3D models for the designed parts by various 3D CAD software such as Pro/E, SolidWorks, SolidEdge, CoreDRAW, UG, I-DEAS, etc.

(2) Generate data conversion file.

Convert the obtained 3D surface or solid model to the data file. It can be formats as STL, IGES, etc. Among them, STL format is the most common one.

(3) Layering.

The 3D model is discretized into a series of ordered 2D slices along a certain direction (usually the height direction). For example, a cone can be seen as a series of planes superimposed in a certain order.

(4) Information processing for each layer.

According to the contour information of each layer, process planning is carried out, processing parameters are selected, and NC code is automatically generated.

(5) Layer processing and stacking.

The forming machine makes a series of thin sheets and automatically stacks them to obtain three-dimensional physical entities.

(6) Post-processing.

Clean the part surface and remove the auxiliary support structure.

Typical RP manufacturing processes include the light curing solid modeling technique, layered solid manufacturing technique, selective sintering technique, melt deposition manufacturing technique, 3D printing and forming technique, etc. The main raw materials can be paper, wax, plastic, photosensitive resin, metal, etc. The forming technology includes photocuring and spray forming.

3) Main methods of rapid prototype manufacturing

(1) SLA (stereolithography apparatus) method.

SLA method is the earliest rapid prototyping technology, and also the most mature technology. It is also called photocuring forming process or laser stereo lithography process. The process principle is that under the irradiation of a certain amount of laser or ultraviolet light, the liquid photosensitive resin undergoes rapid photopolymerization reaction, the molecular weight increases sharply, and the material changes from liquid to solid.

Figure 5-7 shows the schematic diagram of the stereo printing process. There is liquid photosensitive resin (also known as liquid light-curing resin) in the liquid tank. The laser beam can scan the liquid surface under the action of the deflection mirror. The scanning track and the presence or absence of light are controlled by the computer. The photosensitive material will be solidified where the laser irradiates. At the beginning of forming, the worktable is located at a certain depth below the liquid level, and the laser scans and cures point by point on the liquid level according to the instructions of the computer. After the first layer of scanning is completed, the place where the laser is irradiated becomes solid, and the place not irradiated is still liquid. Then, the lift table moves downward by one layer, and the newly formed solid is submerged in the liquid. The scraper scrapes the resin surface with high viscosity, and then the laser starts to irradiate again point by point, curing point by point, until the second layer is completed. Repeat the above process until the whole part is manufactured, then, the 3D solid model is obtained. After that, peeling, repairing, polishing, and other processes need to be carried out.

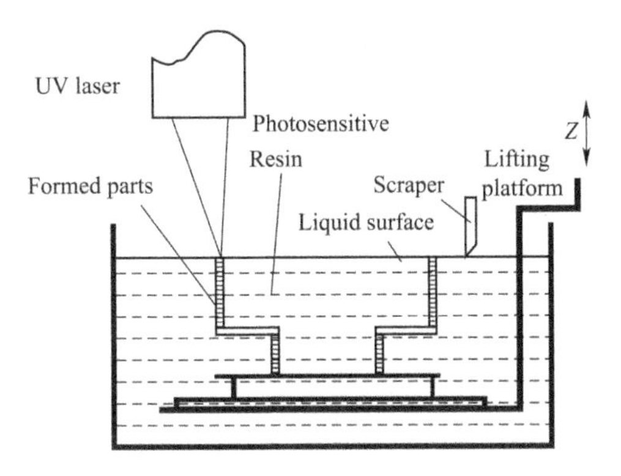

Figure 5-7 Schematic diagram of stereo printing process

The characteristics of SLA are as follows: high manufacturing accuracy, which can achieve to ± 0.10mm, and is independent of the complexity of the workpiece; strong forming ability, which can form fine structures, buckles, and decorative lines; realistic post-processing effect, which is mainly due to the low hardness of photosensitive resin; easy to polish and modify, and good surface finish of the workpiece itself; the strength of the material is slightly worse than ABS; not temperature resistant.

(2) LOM(layered object manufacturing).

LOM is also known as layered solid manufacturing or layered solid manufacturing. The process principle is shown in Figure 5-8.

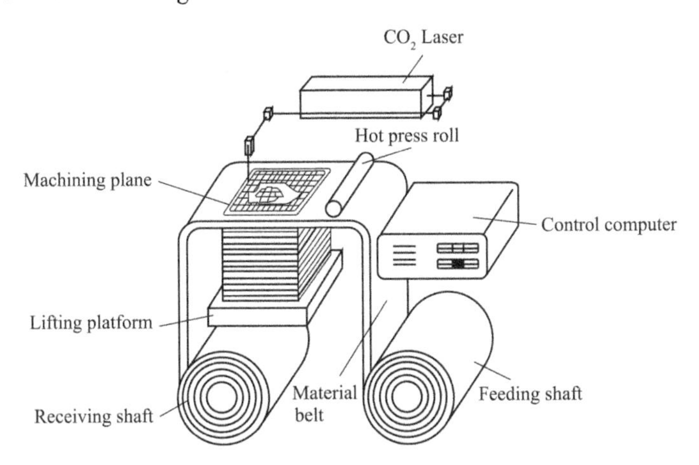

Figure 5-8 LOM process schematic diagram

The LOM process uses thin materials like paper, plastic film, metal foil, etc. The surface of the sheet is coated with a layer of hot melt adhesive in advance. During processing, the laser will cut the sheet material on the worktable according to the layered profile information intercepted by the slicing software, and then use hot rolling to press the sheet material to make it bond with the formed workpiece below; use CO_2 laser cuts the part section profile and the outer frame of workpiece on the newly bonded layer, and cuts the mesh in the redundant area between the section profile and the outer frame, this is convenient for separation of post-processing. After the laser has finished one section layer, the worktable drives the formed workpiece to drop a layer thickness and separate from the strip sheet material (strip); the feeding mechanism rotates the receiving shaft and the feeding shaft to drive the feeding belt and moves the new layer to the processing area; the worktable rises to the processing plane; the number of layers of the workpiece is increased by one layer, the height is increased by one layer thickness, and then the section profile is cut on the new layer. Repeat the above process until all sections of the part are bonded and cut to obtain the solid parts.

The characteristics of LOM are as follows: simple process, fast forming speed, and suitable for making large parts. The LOM process cut only the profile of the part section on the sheet material instead of scanning the whole section. Therefore, the speed of forming thick wall parts is faster, and it is easy to manufacture large parts. The redundant material between the workpiece frame and the section profile plays a supporting role in the processing, and all LOM processes do not need to be supported. The accuracy of the parts is high, and it can reach 0.1mm for laser cutting and 0.15mm for tool cutting. There is no material phase change during the process, and it is difficult to cause warpage deformation. The raw materials are extensive, low cost, and environmentally friendly when using paper as raw material. The mechanical properties are poor and only suitable for appearance inspection.

(3) SLS(selective laser sintering).

The SLS process was developed by C. R. Dechard of the University of Texas at Austin in 1989. This method has been commercialized by DTM in the United States, and SLS Model 125 forming machine has been launched. German EOS company and China's Beijing Longyuan Automatic Forming System Co., Ltd. also launched their own SLS process forming machines of EOSINT and AFS.

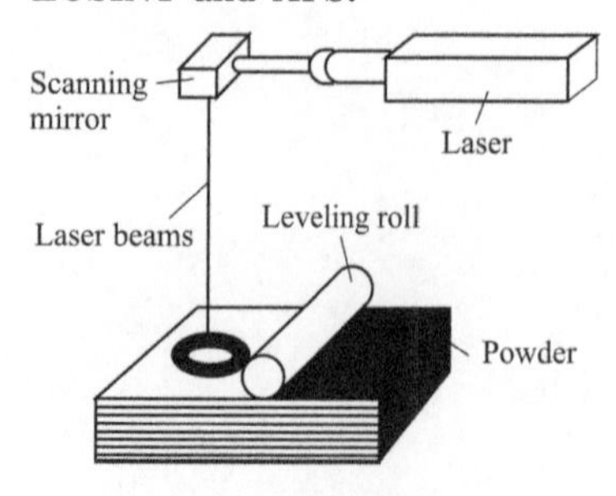

Figure 5-9 SLS process principle

The SLS process is formed by using powder materials. Its principle is shown in Figure 5-9. Spread the material powder on the upper surface of the formed parts and scrape it flat. Use a high-strength CO_2 laser to scan the section of the part on the newly laid layer. The material powder is sintered together under high-intensity laser irradiation to obtain the section of the part and is bonded with the formed part below. Once one layer of the cross-section is sintering, a new layer of material powder is paved, and the lower layer of the cross-section is selectively sintered.

(4) DSPC(direct shell production casting).

DSPC comes from 3D printing rapid prototyping technology. The processing process is as follows: put the part model designed by CAD into the mold shell design device, draw the casting mold shell by microcomputer, generate an electronic model of the mold shell component that has specified thickness and is equipped with a mold core, and transfer it to the mold shell manufacturing device, and then make a 3D ceramic mold shell from the electronic model. Remove the loose ceramic powder from the mold shell, expose the completed mold shell, and finally process the mold shell by the general investment casting method to complete the whole process. This system can detect printing defects and complete all processing without drawing.

4) Rapid prototype manufacturing technical features

(1) Without any tool, mold and fixture, the design scheme of any complex shape can be quickly converted into 3D model or sample, which greatly shortens the production cycle of new products, and the processing efficiency is much higher than that of NC machining.

(2) The model or sample can be directly used for new product design verification, function verification, appearance verification, engineering analysis, market orders, and enterprise decision-making, which is very conducive to early error finding, early modification, and early optimization, improving the first-time success rate of new product development, shortening the development cycle and reducing the R&D cost.

(3) Fast, accurate and complex model manufacturing.

(4) Combining CAD/CAM technology, laser technology, computer numerical control technology, precision servo drive technology, and new material technology.

5) Application of rapid prototyping manufacturing

(1) SLA practical application.

① Application in automobile body manufacturing. SLA technology can produce the required precision proportion casting mold, thus casting a certain proportion of the body metal model. This metal model can be used to conduct wind tunnel and collision tests, to complete the final evaluation of the body to determine whether the design is fair. Chrysler has made a car body model with SLA technology and put it in the high-speed wind tunnel for aerodynamic test analysis, which has achieved satisfactory results and saved the test cost greatly.

② Application in intake pipe test of the automobile engine. The shape of the inner cavity of the intake pipe is made of complex free surfaces, which have a great impact on the intake efficiency and combustion process. During the design process, it is necessary to conduct the air passage test for different air inlet pipe schemes. The traditional method is to manually process the wood mold or gypsum mold of the air inlet pipe described by dozens of sections and then cast the air inlet pipe with sand mold. During the processing, the misunderstanding of the drawings and the technical limitations of the wood mold workers often lead to deviation from the design, and sometimes the impact of this error is obvious. Although NC machining can reflect the design intent better, its preparation time is long, especially when the geometric shape is complex. The British Rover Company used rapid prototyping technology to produce the external mold and internal cavity mold of the intake pipe and achieved satisfactory results, as shown in Figure 5-10.

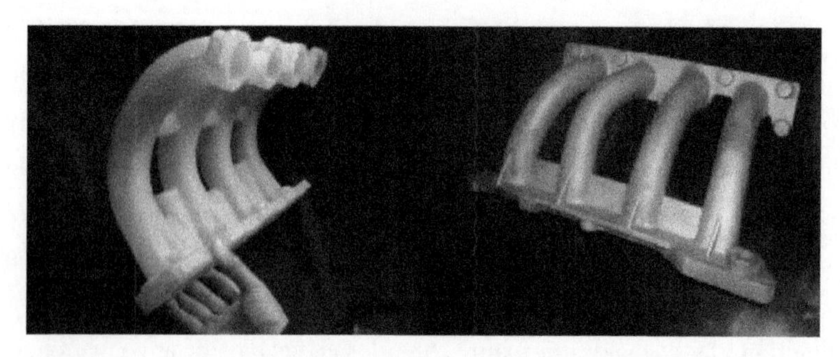

Figure 5-10 Automobile engine intake pipe and actual metal parts made with SLA

③ Application in the aerospace field. SLA technology has no thermal effect, and it can be made into complex and fine parts of various sizes and specifications, with a wide range of applications and good comprehensive stability. It is the only rapid prototyping technology that can meet the accuracy, surface quality, and stability requirements of the aerospace products.

④ Application in electronic products. For example, computers and peripheral products, such as audio, cameras, mobile phones, MP3 players, palmtop computers, cameras, etc. , as well as some household appliances with complex structures, such as electric irons, hair dryers, vacuum cleaners, etc. ,shown in Figure 5-11.

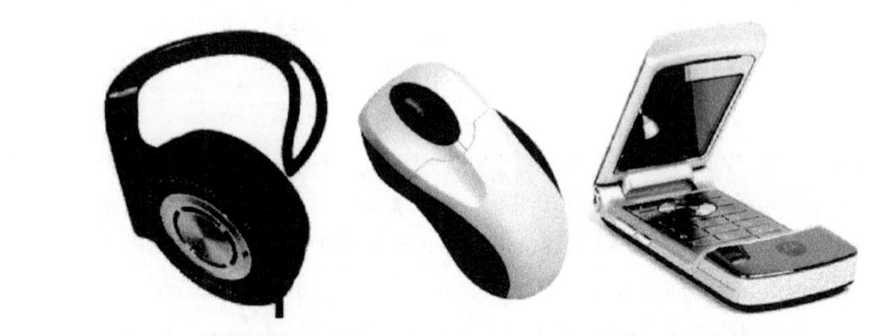

Figure 5-11 SLA headset, mouse, and mobile phone

Figure 5-12 Automobile lamp made by LOM process

(2) LOM practical application.

LOM has a wide range of applications, particularly in automobile lamps, shoemaking, moulds, etc. Figure 5-12 shows the automobile lamps made by the LOM process.

(3) SLS practical application.

SLS technology has been successfully applied in many industries, such as automobile, shipbuilding, aerospace, communications, MEMS, architecture, medical treatment, archaeology, etc. , which has injected new creativity into many traditional manufacturing industries and also brought the informatization. The SLS process can be applied to the following situations.

① Rapid prototype manufacturing. The SLS process can rapidly manufacture the prototype of the designed parts and timely evaluate and correct the products to improve the design quality. It can enables customers to obtain intuitive part models. It can make complex models for teaching and experimenting.

② Preparation and development of new materials. The SLS process can be used to develop some new particles to strengthen composites and cemented carbide.

③ Manufacturing and processing of small batches and special parts. It often encounters the situation of producing small batches and particular complex parts. These parts have a long processing cycle and high cost and even cannot be manufactured. SLS technology can economically realize the manufacturing of small batches and complex parts.

④ Rapid tooling and tool manufacturing. Parts manufactured by SLS can be directly used as molds, such as investment casting, sand casting, high-precision metal models with complex shapes, etc. Formed parts can also be as the functional parts utilized after post-processing. Figure 5-13 is the aircraft model manufactured by the SLS process.

Figure 5-13 The aircraft model manufactured by the SLS process

⑤ Application in reverse engineering. The SLS process can reconstruct the prototype CAD model without design drawings or with incomplete drawings, by the existing part prototype, and by utilizing various digital technologies and CAD technologies. Figure 5-14 demonstrates the reverse engineering implementation process.

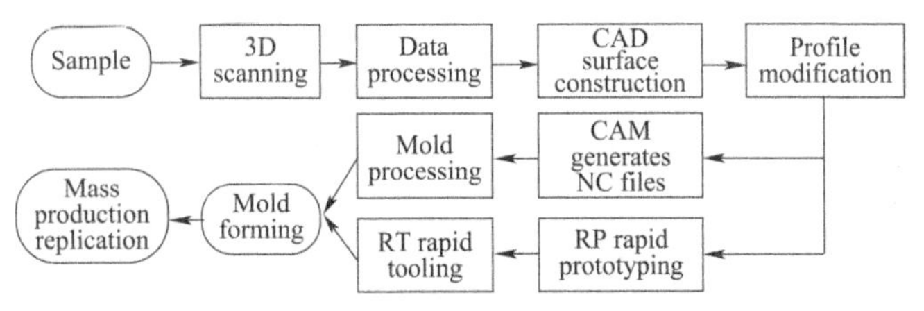

Figure 5-14 Reverse engineering implementation process

5.2.4 Laser Processing

1) Concept and Development

Laser is one of the great inventions of mankind in the 20th century. It has been widely used in industry, military, scientific research, and daily life. In the 21st century, it is claimed that mankind has entered the era of photoelectron. The further extensive application of laser technology as an energy photoelectron will greatly change human production and life. Laser processing technology, which combines optical, mechanical, and electrical technologies, is an advanced manufacturing technology currently at active penetration into many processes in traditional manufacturing technology. It has been widely used in many fields, such as automobile, metallurgy, aerospace, machinery, daily necessities, and industrial supplies manufacturing, due to its non-contact, no need for molds, cleaning, high efficiency, convenient implementation of numerical control, and special processing. It is used for drilling, cutting, milling, material surface modification, and material synthesis.

2) Basic principle and characteristics of laser processing

When the laser is focused on the material, the phenomenon of temperature rise, heating, melting, or evaporation will occur. The laser beam can be used to cut, connect, pierce, and quench metal materials such as steel plates, plastic, and cloth. In addition, laser can improve the mechanical properties of the metal surface. This process is called laser spray hardening.

The diameter of the laser spot varies from 0.1 ~ 0.3mm, and it is high-density energy, always used for high-precision and ultra-fine machining with a high melting point. Generally, the work table of the laser processing machine is controlled by a program and can process workpieces with complex shapes. Moreover, the laser processing machine belongs to non-contact processing, and its output control is easy and suitable for the automatic control system.

The laser processing's characteristics include no machining tools, no tool consumption, no machining deformation, fast machining speed, small heat affected zone.

By adjusting the beam energy, spot diameter and beam moving speed, various processing can be realized, including microprocessing and automatic processing.

The high power of the laser can process almost all fusible and non-decomposable metal and non-metallic materials, and even transparent materials (such as glass).

The workpiece can be handled through transparent media, such as glass, emotional gas or air, which is very important and convenient in some special cases, such as welding in vacuum tubes.

Laser is easy to guide, focus and diverge, and can be combined with CNC machine tools and robots to form a variety of flexible processing systems, which is conducive to the transformation of traditional processing technology, machine tools and equipment.

Disadvantages of the laser processing: Laser beam processing is a kind of thermal processing under many influencing factors, so it is difficult to guarantee and improve the processing accuracy. Laser harms human body, thus, the corresponding protective measures must be taken.

3) Application of laser processing

(1) Laser drilling.

Laser drilling is the high-energy laser irradiation on the surface of the workpiece, resulting in a series of thermophysical phenomena on the surface, to form holes. Laser drilling technology is widely used in industrial production due to its advantages of high speed, high efficiency, good economic benefits, and wide application fields. Laser can be used for punching and cutting on textile fabrics, leather products, rubber products, paper products, metal products, and plastic products. The application fields include clothing, shoes, handicraft production, machinery and equipment, parts production, and printed circuit board, etc. In addition, laser drilling is also widely applied in diamond molds, watches, jewel bearings, and ceramics. Special holes can also be made by using different light spot shapes, as shown in Figure 5-15.

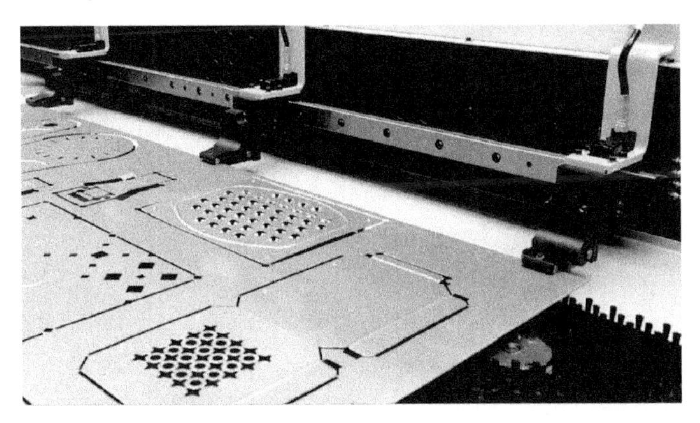

Figure 5-15 Laser drilling parts

(2) Laser cutting.

The working principle of laser cutting is similar to that of laser drilling, but the workpiece and laser beam must move relatively. In actual processing, laser numerical control cutting can be realized using the workbench numerical control technology. Laser cutting uses a focused high-power density laser beam to irradiate the workpiece so that the irradiated material can rapidly melt, vaporize, ablate, or reach the ignition point. At the same time, the molten material can be blown away by high-speed airflow coaxial with the beam to realize the cutting of the workpiece, as shown in Figure 5-16. Laser can cut metal and nonmetal. The laser cutting process does not produce any mechanical impact and pressure on the material, and its cutting seam is small, which is convenient for automatic control. Therefore, in practice, it is often used to process glass, ceramics, and various precise small parts. Laser cutting is one of the most prevalent cutting methods.

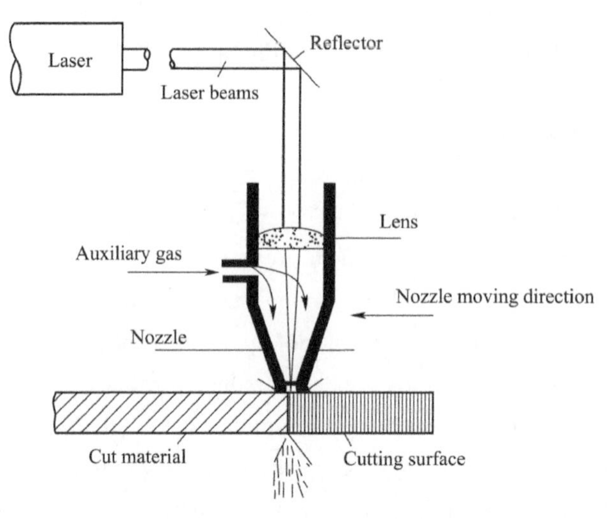

Figure 5-16 Laser cutting principle

In automobile manufacturing, the cutting technology of space curves, such as car roof window, has been widely used. German Volkswagen uses a 500W laser to cut complicated body sheets and various curved parts. In the aerospace industry, laser cutting technology is mainly used for cutting special aviation materials, such as titanium alloy, aluminum alloy, nickel alloy, chromium alloy, stainless steel, beryllium oxide, composite materials. Aerospace parts and components processed by laser cutting include engine flame tube, titanium alloy thin-walled casing, aircraft frame, titanium alloy skin, wing stringer, tail wall panel, helicopter main rotor, space shuttle ceramic thermal insulation tile, etc. Laser-cutting forming technology is also widely used in the field of non-metallic materials. It can cut high hardness and brittleness materials, such as silicon nitride, ceramics, quartz, etc. , and flexible materials, such as cloth, paper, plastic sheet, rubber, etc. 10% to 12% material can be saved and the efficiency can be improved more than 3 times by the laser. Figure 5-17 is the laser cutting robot L-1000 CO_2. Laser cutting parts are shown in Figure 5-18.

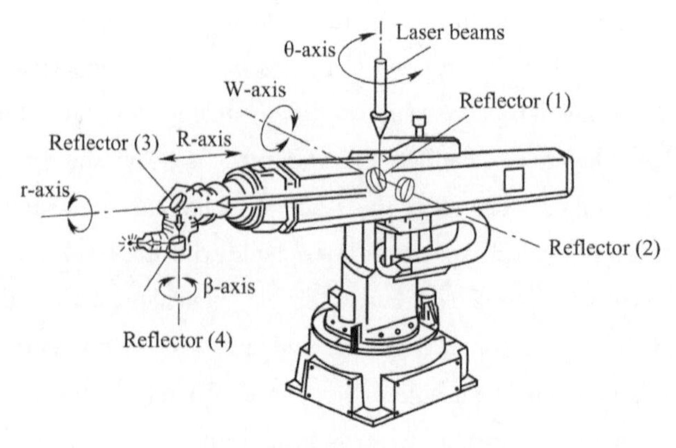

Figure 5-17 L-1000 CO_2 laser cutting robot

Figure 5-18 Laser cutting parts

(3) Laser welding.

Laser welding is to radiate a high-intensity laser beam to the metal surface. Through the interaction between the laser and the metal, the metal absorbs the laser and converts it into heat energy, the metal is melt by the energy and then cool and crystallize to welding, as shown in Figure 5-19.

Laser welding is widely used in the following industries.

① In the manufacturing industry.

Laser tailor-welding technology is widely used in foreign car manufacturing. Research on ultra-thin plate welding, such as foil plate with thickness below 100 μm, cannot be fused, but it can be successfully welded by YAG laser with special output power waveform, which shows the broad prospect of laser welding. The laser tailor-welded car body is shown in Figure 5-20.

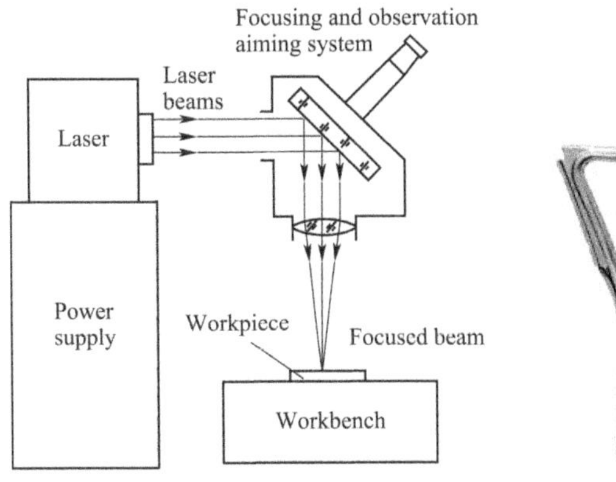

Figure 5-19 Laser welding principle

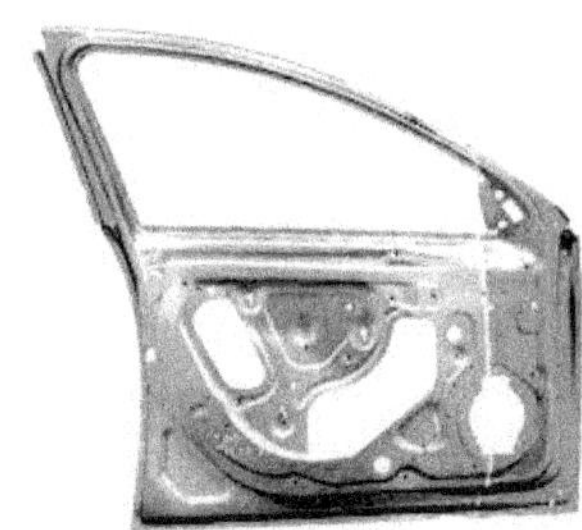

Figure 5-20 Laser tailor-welded car body

② In the powder metallurgy industry.

With the continuous development of science and technology, many industrial technologies have special requirements for materials, and the materials manufactured by smelting and casting methods can no longer meet the needs. Due to the special properties and manufacturing advantages of powder metallurgy materials, it is replacing the traditional smelting and casting materials in some fields, such as automobile, aircraft, tool and cutting tool manufacturing. But the connection between powder metallurgy materials and other parts is a serious problem, which limits its application. In the early 1980s, laser welding entered the powder metallurgy material processing with its unique advantages, opening up a new application prospect for the powder metallurgy materials. For instance, diamond can be welded by the brazing method which is commonly used in the connection of powder metallurgy materials. Laser welding can improve welding strength and high-temperature resistance.

③ In the field of automobile industry.

Laser welding production lines have appeared in the automobile manufacturing industry on a large scale and become one of the outstanding achievements of the automobile manufacturing industry. European automobile manufacturers such as Audi, Mercedes-Benz, and Volkswagen from Germany took the lead in using laser welding for the roof, body, side frame, and other metal welding as early as the 1980s. In the 1990s, General Motors, Ford, and Chrysler in the United States competed to introduce laser welding into automobile manufacturing. Italy Fiat has adopted laser welding in the welding and assembly of most steel plate components. Japan's Nissan and Honda have used laser welding and cutting processes to manufacture body panels. High-strength steel laser welding assemblies are increasingly used for manufacturing automobile bodies due to their excellent performance.

5.2.5 Green Manufacturing

1) Concept and development

Green manufacturing is also known as environmentally-aware manufacturing and environment-oriented manufacturing. In the process of machinery manufacturing, environmental factors are taken into account. Its purpose is to use technical means to optimize the manufacturing process, reduce environmental pollution, and achieve the goal of saving resources and sustainable development. Modern machinery manufacturing includes different stages such as product design, product manufacturing, packaging, transportation, and product recycling. Green manufacturing is the industrial upgrading and optimization of the machinery manufacturing industry through improving the resource recycling rate, optimizing resource allocation, and reasonably protecting the environment.

The research on green manufacturing and its related issues has been very active in recent years. Especially in the United States, Canada, Western Europe, and other developed countries, much research has been carried out on green manufacturing and related issues. In China, a lot of research has been carried out on green manufacturing and related issues in recent years.

2) Main factors of green manufacturing technology

The basic mode of green manufacturing is divided into three aspects: green resources, green production, and green products.

(1) Green resources.

Green resources refer to using green materials and energy in the manufacturing process. From the perspective of raw materials and energy, trying to select some materials with less pollution or no pollution, and these materials should have the value of recycling to meet the basic requirements of green manufacturing.

(2) Green production.

Green production includes green design and green production process. Green design should fully considerate manufacturing products, such as product quality, product life, and the relationship between products and the environment. In addition, environmental factors should also be effectively considered in the design process. The green production process refers to effectively reducing environmental pollution and saving energy in product production. For the rational use of green processes, it is necessary to think of the technical aspects of energy conservation and environmental protection, further carry out detailed research on how to reduce the pollution of products, and fully recognize how to reduce the materials, how to reduce the waste of materials, and how to reduce the impact on the environment.

(3) Green products.

Green products refer to reducing energy consumption and environmental pollution in the later cycle of products by combining the green concept in the process of packaging, transportation, use, and recycling of products.

3) Characteristics of green manufacturing

Green manufacturing has the characteristics of comprehensiveness, integrity, and intersection. Comprehensiveness means that green manufacturing must run through the entire product life cycle. It covers every stage of the product life cycle and requires different measures for different stages. Integrity refers to the impact on the environment at different stages of product manufacturing. Green manufacturing should not only consider the specific causes of environmental damage but also consider the relationship between resources, equipment and products, and should grasp the internal relationship as a whole to achieve the goal of optimizing, upgrading, and reducing pollution. Intersection means that the whole green manufacturing process covers a variety of disciplines, such as mechanical manufacturing technology, material technology, environmental management, etc., reflecting the interdisciplinary characteristics.

4) Development trend of green manufacturing

At present, there is a "green wave" in the world. Environmental issues have become the focus of all countries worldwide and have been listed on the world development agenda. The manufacturing industry will change the traditional manufacturing mode, promote green manufacturing technology, develop relevant green materials, green energy, green design database, knowledge base, and other basic technologies, and produce green products that protect the environment and improve resource efficiency, such as green cars, green refrigerators, etc, and regulate enterprise behavior with laws and regulations. With the enhancement of people's awareness of environmental protection, those enterprises that do not implement green manufacturing technology and do not produce green products will be eliminated from the market competition, making it imperative to develop green manufacturing technology.

(1) Globalization.

The global characteristics of green manufacturing are reflected in the following aspects: the impact of manufacturing on the environment is often not limited in the local area, and environment protection needs global unity. With the formation of the global market in recent years, the market competition for green products will be global. In recent years, many countries have required imported products to be identified as green and have "green marks". Green manufacturing will provide technical means for Chinese enterprises to improve the greenness of products, thus providing strong support for Chinese enterprises to eliminate international trade barriers and enter the international market.

(2) Socialization.

The research and implementation of green manufacturing need the joint efforts and participation of the whole society to establish the necessary social support system for green manufacturing. The social support system involved in green manufacturing is the issue of legislation and administrative regulations. Secondly, the government can formulate economic policies and use the mechanism of the market economy to guide green manufacturing. Using economic means to strictly control non-renewable resources and resources that are renewable but have an impact on the environment which can force enterprises to reduce the direct utilization of such resources as much as possible and develop alternative resources. To truly and effectively implement green manufacturing, enterprises must consider the treatment of the end product, which may lead to a new integrated relationship between enterprises, products, and users.

(3) Integratedization.

Green manufacturing involves the whole process of the product life cycle and all aspects of the production and operation activities of the enterprise, it is a complex system problem. At present, the integrated functional objective system of green manufacturing, the integration of product, process design and material selection, the integration of user needs and product use, the integration of information systems in green manufacturing, and the integration of green manufacturing processes will become important research topics on green manufacturing.

(4) Parallelization.

An significant trend of green design in the future is to combine with concurrent engineering to form a new product design model. It is green concurrent engineering. It is a systematic method that can design products in an integrated and parallel way and makes product developers consider the whole life cycle of products at the beginning of design.

(5) Intelligentization.

Intelligence is the combination of artificial intelligence and intelligent manufacturing technology. The decision objective system of green manufacturing is the integration of the existing manufacturing system objective system with environmental impact and resource consumption. To optimize these objectives, it needs to use artificial intelligence methods to support processing. In addition, the green product evaluation index system and evaluation expert system both need artificial intelligence and intelligent manufacturing technology. Therefore, artificial intelligence technology based on knowledge system, fuzzy system, and neural network will play an essential role in the research and development of green manufacturing.

5.3 Manufacturing Automation Technology

Manufacturers are seeing an increasing need for automation year over year. The manufacturing industry is one of the most prevalent and important areas for the use of automation technology. As more manufacturers look for ways to drive efficiency and lower costs, automation in factories, Industry 4.0, and other connected manufacturing solutions will continue to become more widespread.

In the context of manufacturing, automation is the use of equipment to automate systems or processes. The end goal of manufacturing automation is to increase production capacity while reducing costs. Electromechanical systems can be programmed to perform a variety of tasks, and automation is particularly helpful for repetitive tasks or tasks that require extreme precision. Automation can also be used in manufacturing business management, such as with automated inventory scheduling, and sending and analyzing data for reporting, and can help improve workplace safety by performing tasks that could injure or endanger human workers.

5.3.1 Definition and Connotation of Manufacturing Automation

The early concept of manufacturing automation refers to transferring parts between machines without human handling in the production process. The concept of manufacturing automation (hereinafter referred to as automation) is a dynamic development process. With the development of electronic and information technology, especially the development and wide application of computers, the concept of automation has been expanded to use machines (including computers) not only to replace human physical labor but also to replace or assist mental labor to complete specific tasks automatically.

The broad connotation of manufacturing automation includes the following points:

(1) Form.

Manufacturing automation has three meanings: replacing human physical labor, replacing or assisting people's mental work, coordination, management, control, and optimization of labor, machine and the whole system in the manufacturing system.

(2) Function.

The functional objectives of manufacturing automation are multifaceted and can be described using TQCSE model. T presents the production time and productivity of the product, Q presents the quality of the product, C presents the cost of production, S presents the product service and manufacturing service, and E presents the environment.

(3) Scope.

It involves all processes of the product life cycle, not only specific production and manufacturing processes.

5.3.2 The Development Process and Trend of Manufacturing Automation

1) The development process of manufacturing automation

The history and development of manufacturing automation can be divided into four stages, as shown in Table 5-1.

Manufacturing automation technology development process Table 5-1

Stage of development		Name	New technologies introduced	Characteristic	Background of mechanical manufacturing systems and related technology	Time of development and application	Scope of application
Stage I	Traditional mechanical manufacturing automation	Automatic single machine production line	Relay program control, modular machine tool	High efficiency, high rigidity	Traditional mechanical design and manufacturing process method	1940s—1950s	Mass production
Stage II	Modern machinery manufacturing automation	CNC machine tool machining center	NC/CNC	Flexibility, process concentration	Electronic technology/ digital circuit computer programming technology	1950s - 1970s (NC) 1970s - 1980s(CNC)	Singleton, small batch, multivariety

Continue

Stage of development		Name	New technologies introduced	Characteristic	Background of mechanical manufacturing systems and related technology	Time of development and application	Scope of application
Stage Ⅲ	Modern machinery manufacturing automation	Flexible manufacturing system, flexible production line	CAD/Industrial robot/CAM/group technology/DNC/FMS/FML	Ideal combination of flexibility and efficiency	Computer geometry technology, discrete event system theory method and simulation technology; workshop production planning and control; computer control and communication network	1970s – 1980s	Medium and small batches; multivarity; mass production
Stage Ⅳ		Computer integration, manufacturing system	CAD/CAPP/CAM integrated production management and scheduling; information technology of automatic processing system; simulation technology and workshop dynamic scheduling	Fully automated, optimized, intelligent and networked information processing	Integrated information processing of design, process, planning and manufacturing; AI/intelligent manufacturing organization, decision support, distributed network communication and data resource sharing throughout the plant	After 1980s	Automation of design, manufacturing and economic management

(1) Rigid automation. Rigid automation includes rigid automatic line and automatic single machine. This stage matured in the 1940s and 1950s. Apply the traditional mechanical design

and manufacturing process methods, mass production is achieved by special machine tools, modular machine tools, automatic single machine or automatic production line. It is characterized by high productivity and rigid structure, but it is difficult in changing products. The new technologies introduced include relay program control, modular machine tool, etc.

(2) CNC machining. CNC machining includes numerical control (NC) and computer numerical control (CNC). Numerical control (NC) in this stage has developed rapidly and matured in the 1950s and 1970s. It was rapidly replaced by computer numerical control (CNC) in the 1970s and 1980s due to the rapid development of computer technology. CNC machining equipment includes CNC machine tools, machining centers, etc. CNC machining is characterized by good flexibility and high processing quality and is applicable for producing multivarities and small /medium-sized batches (including single products). The new technologies introduced include numerical control, computer programming etc.

(3) Flexible manufacturing. Flexibal manufacturing emphasizes the flexibility and efficiency of the manufacturing process. It is applicable to the production of multispecies and small and medium-sized batches. The main technologies involved include group technology (GT), computer direct numerical control and distributed numerical control (DNC), flexible manufacturing cell (FMC), flexible manufacturing system (FMS), flexible manufacturing line (FML), discrete system theory and method, simulation technology, workshop planning and control, manufacturing process monitoring technology, computer control, and communication network, etc.

(4) Computer Integrated Manufacturing System (CIMS). CIMS can be seen as a new stage in the development of manufacturing automation, and also as a higher-level manufacturing automation system. CIMS has been developing rapidly since the 1980s and is now in the ascendant. Its characteristic is to emphasize the systematization and integration of the whole manufacturing process to solve the TQCS problems of the survival and competition of modern enterprises, that is, shorter time to market, higher quality, lower cost, and better service. CIMS involves a wide range of disciplines and technologies, including modern manufacturing technology, management technology, computer technology, information technology, automation technology, and system engineering technology.

(5) New manufacturing automation models. New manufacturing automation models include intelligent manufacturing, agile manufacturing, virtual manufacturing, network manufacturing, global manufacturing, green manufacturing, etc. The new manufacturing automation model mentioned above was proposed and studied at the end of the 20th century, which is the development direction of manufacturing automation in the 21st century.

2) The development trend of manufacturing automation

(1) Intelligent. Intelligent manufacturing system is a human-machine integrated intelligent system composed of intelligent machines and human experts. It carries out intelligent activities such as analysis, reasoning, judgment, conception and decision-making in the manufacturing

process. The goal of intelligent manufacturing technology is to expand and replace human mental labor through the cooperation between intelligent machines and people.

(2) Agility. Agile manufacturing is a manufacturing strategy and modern manufacturing mode facing the 21st century, which mainly involves the agility of the manufacturing environment and manufacturing process.

(3) Flexibility. Flexibility includes machine flexibility, process flexibility, operation flexibility and expansion flexibility.

(4) Reconstruction ability. It can realize rapid reorganization and reconstruction, and enhance the rapid response ability to new product development.

(5) Rapid integrated manufacturing process, such as rapid prototype manufacturing RPM.

(6) Globalization. The concept of manufacturing globalization stems from the intelligent system plan of developed countries such as the United States, Japan and Europe. In recent years, with the development of internet technology, the research and application of manufacturing globalization have developed rapidly. Manufacturing globalization covers a wide range of contents, mainly including the internationalization of the market, the formation of the global network of product sales, the international cooperation of product design and development, the transnational of product manufacturing, the restructuring and integration of manufacturing enterprises worldwide, the cross-regional and cross-national coordination, sharing and optimization of manufacturing resources, and the formation of the global manufacturing system structure.

(7) Networking. At present, due to the rapid development of network technology, especially Internet/Intranet technology, it is bringing new changes to enterprise manufacturing activities. The depth, breadth, and development speed of its impact are often far beyond people's expectations. Network-based manufacturing includes the following aspects: networking within the manufacturing environment to realize the integration of manufacturing processes. The manufacturing environment is networked with the entire manufacturing enterprise to realize the integration of the manufacturing environment with the engineering design, management information system, and other subsystems in the enterprise. Networking between enterprises to realize resource sharing, combination and optimal utilization among enterprises. Realize remote manufacturing through the network.

(8) Virtualization. Virtual manufacturing is a comprehensive technology based on system modeling technology and simulation technology supported by manufacturing technology and computer technology, which integrates modern manufacturing technology, computer graphics, concurrent engineering, artificial intelligence, artificial reality technology, and multimedia technology. It is formed by multidisciplinary knowledge and is a key technology to realize agile manufacturing. In the future, virtual manufacturing will develop into a large software industry.

(9) Green manufacturing. Its goal is to minimize the impact on the environment and maximize the resource efficiency in the entire product life cycle from design, manufacturing,

packaging, transportation, and use to scrap disposal. Green manufacturing is the embodiment of sustainable development strategy in the manufacturing industry, also, it is the sustainable development mode of the modern manufacturing industry.

5.3.3 Key Technologies for Manufacturing Automation

(1) Integration technology and system technology in manufacturing system have become hot issues in manufacturing automation research.

In the past, the research of manufacturing automation technology mainly focused on the research of unit and specialized technology, including control technology, such as numerical control technology, process control and process monitoring, and computer-aided technology, such as CAD, CAPP, CAM, and CAE. In recent years, while the technologies mentioned continue to be improved, the research on integration and system technology in manufacturing systems has developed rapidly. It has become a hot topic in manufacturing automation research. Integration technology and system technology in manufacturing systems cover a wide range. The integration technology includes information integration and function integration technology, such as CIMS; process integration technology, such as CE; inter-enterprise integration technology, such as agile manufacturing AM, etc. System technology includes manufacturing system analysis technology, manufacturing system modeling technology, manufacturing system operation research technology, manufacturing system management technology, and manufacturing system optimization technology.

(2) Pay more attention to the role of humans in the research of manufacturing automation system.

In the past, people used to think that complete automation and unmanned chemical plants or workshops were the development goals of manufacturing automation. With the deepening of practice and the failure of the implementation of some unmanned chemical plants, people have reflected on the problem of unmanned manufacturing automation and have a new understanding of the irreplaceable important role of humans in the manufacturing automation system. Therefore, many countries have launched active exploration in theory and technology on organically integrating human and manufacturing systems. Over the past few years, new ideas such as "human-machine integrated manufacturing system" and "human-centered manufacturing system" have been put forward. Its connotation is that take humans as an organic part of the system structure, put humans and machines in an optimized cooperation position, and realize the human-machine integrated decision-making mechanism in the manufacturing system to achieve the best benefits of the manufacturing system. At present, a lot of research is being carried out around human-machine integration.

(3) The research of unit system still occupies an important position.

The unit system is the general name of the mechanical processing system, which is based on one or more numerical control processing equipment and material storage and transportation systems and can carry out the automatic processing and production of multispecies, small and medium-sized batches of parts under the unified control and management of the computer. It is an integral part of the CIMS (Computer Integrated Manufacturing System). It is the decomposition decision-making level and specific execution organization of the work plan for the automation workshop. The manufacturing industry has invested many manual labor and material resources in the theoretical and technical research of unit systems, so unit technology has developed rapidly in software and hardware.

(4) The manufacturing process planning and scheduling research is very active, but the practical results are still rare.

Ingersoll Milling Machine Company of the United States once analyzed the manufacturing process from raw materials to products in traditional manufacturing plants. The results show that for a mechanical part, only 5% time is spent on the machine tool; 95% is spent on transporting or waiting between different places and machine tools. Reducing this 95% time is an important direction to improve manufacturing productivity. Optimizing the planning and scheduling of the manufacturing process is the primary means to reduce time. Therefore, manufacturing process planning and scheduling research is very active worldwide. However, due to the complexity and randomness of manufacturing processes, few research results are applied to practical applications, especially those with extensive applicability. More research needs to be further deepened.

(5) The research of flexible manufacturing technology is developing in a broad sense.

The current research focuses on the system structure, control, management, and optimal operation of FMS. FMS does not only pay attention to the integration of information flow but also emphasizes the integration and automation of material flow. Hence, the investment in logistics automation equipment accounts for a large proportion of the investment in the entire FMS, and the operation reliability of FMS largely depends on the regular operation of logistics automation equipment.

5.4 Advanced Manufacturing Mode

The manufacturing model is an effective production mode and a certain form of production organization that can improve product quality, market competitiveness, production scale, and production speed. It can be adopted by the manufacturing industry to complete specific production tasks. Manufacturing model relates to the enterprise system, production organization form, governance mode, technology system, and operation pattern.

Advanced manufacturing mode refers to the advanced production method that can achieve good manufacturing effects by effectively organizing various manufacturing elements according to different manufacturing environments in the production and manufacturing process.

5.4.1 Lean Production

1) Concept of lean production

Lean production (LP) is a new type of production mode that can maximize benefits to the enterprise. It is achieved by using modern advanced manufacturing technology and management technology, allocating and using enterprise resources effectively and reasonably, optimizing and combining the various elements of the whole process of product formation, eliminating excessive production, and improving product quality.

2) Characteristics of lean production

(1) Take users as "God". Change from "product-centered" to "user-centered". Provide users with products that have suitable prices, excellent quality, and high-quality services.

(2) Focus on "people". Satisfy employees' desire to learn new knowledge and realize their self-worth and carry out regular training.

(3) Take simplification as the approach. Simplify the organizational structure of the enterprise, simplify work processes, adopt flexible processing equipment to reduce direct labor, and reduce inventory to achieve "zero inventory".

(4) Work in parallel. Product development in the form of a work team.

(5) Adopt a group of assembly lines.

(6) Supply appropriate parts and components in a timely and appropriate manner.

(7) Pursue "perfection" and "zero defect".

5.4.2 Networked Manufacturing

Networked manufacturing takes advantage of internet, organizes social manufacturing resources flexibly and rapidly, and combines the existing production equipment resources, intellectual resources, and various core capabilities scattered in different regions to quickly launch new high-quality, low-cost products.

1) Characteristics of networked manufacturing

The networked manufacturing characteristics system is shown in Figure 5-21.

(1) Basic characteristics.

Based on the advanced manufacturing mode of the network, enterprises organize and manage through the Internet, intranet and extranet; respond to market demand quickly; integrate resources and save costs. It can break through geographical and time constraints.

(2) Technical characteristics.

① Time characteristics. The network enables the rapid transmission and interaction of information, which improves the time efficiency of the information transmission process in the manufacturing system dramatically. It makes significant changes in the time process of manufacturing activities, and even produces some special functions. For example, it can make the jet leg in the manufacturing process to work 24 hours without interruption.

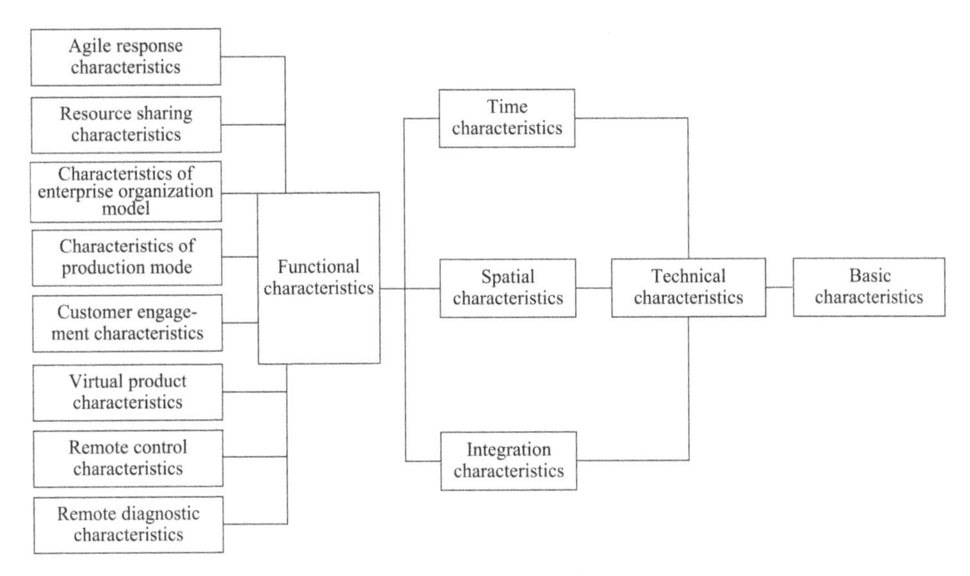

Figure 5-21 Networked manufacturing characteristics system

② Spatial characteristics. The network expands the enterprise space, and the network-based remote design and manufacturing enable enterprises to go out to the world. Also, it enables enterprises scattered around the world to form dynamic alliances at any time concerning market opportunities, realize resource sharing, and form a virtual enterprise with wide space and dynamic changes.

③ Integration features. Rapid transmission and interaction of network and information support information integration, function integration, process integration, resource integration, and integration between enterprises.

(3) Functional characteristics.

① Characteristics of agile response. Technologies such as network-based agile manufacturing and concurrent engineering can significantly shorten the product development cycle and rapidly respond to the market.

② Characteristics of resource sharing. Information resources, equipment resources and even human resources distributed throughout the network can be shared and optimized.

③ Characteristics of enterprise organization mode. Change the traditional enterprise organization mode with strong closeness and pyramidal hierarchical structure to the flat, transparent and project-based organization mode.

④ Characteristics of production mode. Develop the large batch, small variety mode to the small batch, multivarity and customized production mode, which reflects the characteristics of personalized demand. Also, web-based customization will be an effective mode to meet this demand.

⑤ Characteristics of customer participation. Customers are not only consumers of products, but also creators and design participants of products. Network-based DFC and DBC technology provides users with the possibility to participate in product design.

⑥ Characteristics of virtual products. Virtual products, virtual supermarkets, and networked sales will be important ways of market competition in the future. Users can customize, see, and conduct virtual use and performance evaluation of their favorite products at home.

⑦ Characteristics of remote control. Broadband networking operation of equipment can realize remote control and management of equipment and remote sharing of equipment resources.

⑧ Characteristics of remote diagnosis. Remote monitoring and fault diagnosis of equipment and production site can be realized based on the network.

2) Composition of networked manufacturing system

(1) Intelligent public information service platform. Provide intelligent public information services for enterprises, including basic data, basic public information, automatic information collection, classification and matching services.

(2) E-commerce support platform. Build the supply-demand information database of products, technology, talents, and other aspects. Research and develop the customer relationship management system and the agile supply chain management system, and realize the optimized scheduling and management of logistics based on the Internet.

(3) CSCW support environment for the Internet. Study the architecture of the support environment for the networked agile enterprise, develop software support technology and collaboration mechanism based on agile manufacturing methods and theories, and establish a collaborative design environment for distributed and networked product design.

(4) Public data center. Build a manufacturing information resource sharing database on the Internet, which includes enterprise production capacity, special manufacturing means, talent information, various computer-aided software, standard parts, general parts, drawing library, and other technical and product information.

(5) Safety assurance system. Provide a supportive environment and guarantee system for user business security. The enterprise networked manufacturing platform focuses on building its own networked manufacturing platform, and builds a multi-directional, multi-level, and multi-application information platform based on order information flow.

3) Key technologies of networked manufacturing

(1) Integrated technology. Including product lifecycle management, collaborative product commerce, mass customization, and concurrent engineering. The above-integrated technologies include various new management concepts. With the support of corresponding enabling technologies, basic technologies, and supporting technologies combined with the characteristics of networked manufacturing systems, these integrated technologies can effectively solve different problems in networked manufacturing.

(2) Generic technology. Relevant generic technologies are the central technologies of some networked manufacturing systems. Common integration technologies include CAD, CAE,

CAM, CAPP, ERP, SCM, PDM, customer relationship management, supplier relationship management, and manufacturing execution system, etc.

(3) Basic technology. Relevant basic technologies include standardization technology, product modeling technology, and knowledge management technology. Although these technologies generally could not solve specific problems directly, they also play a very important role as the basis for using enabling technologies to solve specific problems.

(4) Support technology. Relevant supporting technologies, such as computer technology and network technology directly affect the efficiency and reliability of networked manufacturing system, they are the infrastructure for implementing networked manufacturing.

The four technologies mentioned above complement each other to support networked manufacturing for enterprises.

4) Application of networked manufacturing

(1) Application in the mold manufacturing industry. For example, the mold networked manufacturing demonstration system in Shenzhen, China, has formed a mold design and manufacturing network system, which combines CAD/CAM technology, virtual design and manufacturing technology, computer network technology, rapid prototyping and post-processing together, can greatly improve manufacturing capacity and labor efficiency and overcome the shortcomings of the traditional mold manufacturing, such as long cycle, high cost, and slow response.

(2) Application in the dynamic alliance of enterprises. For example, France Aerospace, British Aerospace, Germany DASA, and Spain have formed the Airbus Group. The ACE (Airbus concurrent engineering) of Airbus has adopted the non-local paperless design technology like Boeing and implemented concurrent engineering in the development of Airbus series aircraft to compete with Boeing.

(3) Application in the automobile manufacturing industry. The three major automobile companies General Motors, Ford Motor Company, and DaimlerChrysler terminated their parts procurement plans and turned to jointly establish the e-commerce market for parts procurement, reducing the average cost of each transaction from \$ 100 ~ \$ 150 to less than \$ 5, and reducing the manufacturing cost of each vehicle by at least \$ 1200.

Chapter 6 Modern Design Methods and the Applications

6.1 Connotation of Modern Design Methods

The design methodology is a new developing subject. Until now, there is no exact and recognized definition, research object, or category of the design methodology. The development of design methodology has been extremely fast over the past few years, and it has been widely concerned by scholars from all over the world.

Modern design is the deepening, enrichment, and perfection of traditional design. It is driven by market demand, guided by modern design ideas and theories, centered on knowledge acquisition, taking modern technical means as tools, and considering the whole life cycle of products and the compatibility of people, machines, and the environment. With the help of computers, high-precision analysis methods (such as finite element method, etc.) have been widely used in analysis and calculation, which can improve the initiative, scientificity, and accuracy of the design. We can find the best design scheme from many feasible designs, replace the empirical design, approximate calculation by theoretical design, replace the accurate calculation, and replace the general safety life design by optimized design and green design. Human design work has roughly gone through the following stages.

(1) Intuitive design stage.

It is often blindness to design and make tools and machinery by intuitive feeling.

(2) Experiential design stage.

Depending on personal talent and experience, using some basic design calculation theory, with the help of analogy, simulation and trial and error design methods.

(3) Semi-theoretical and semi-empirical design stage.

It strengthens the basic design theory, the design mechanism of various professional mechanical products, and the design research of essential parts so that designers can fully utilize data, charts, and manuals to design. It can reduce the blindness of design and increase rationality.

(4) Modern design stage.

Due to the rapid development of science and technology, the understanding of the objective world is deepening, and the theoretical basis and means needed for design work have made significant progress, especially the development and application of computer technology, which has produced revolutionary mutations in design work.

6.1.1 Research Content of Modern Design Methodology

The design methodology is a comprehensive subject that studies the law of product design, design procedure, thinking, and working methods. Design methodology analyzes the strategic process and design methods from a systematic point of view. Based on summarizing design rules and inspiring creativity, it promotes the comprehensive application of modern design theory, scientific methods, advanced means, and tools in design, which plays an active role in developing new products, transforming old products and improving market competitiveness of products.

1) The aspects of design methodology

(1) Analyze the design process and the tasks of each design stage, and seek the design program conforming to scientific laws. The design process is divided into four stages: design planning (defining design tasks), scheme design, technical design, and construction design. The main tasks and objectives of each stage are defined. On this basis, the process mode of product development is established, and the optimal design and integrated development strategy of the product life cycle are discussed.

(2) Study the logical steps and working principles to solve design problems. The problem-solving steps of system engineering analysis, synthesis, evaluation, and decision-making go through every stage of design so that the problem can be gradually expanded and multiple schemes can be selected.

(3) Emphasize the importance of designers' innovative ability in product design, analyze the law of innovative thinking, study and promote the application of various innovative techniques in design.

(4) Analyze the application of various modern design theories and methods, such as system engineering, creation engineering, value engineering, optimization engineering, similarity engineering, ergonomics, industrial aesthetics, etc. It should realize the scientific and reasonable design of products and improve the competitiveness of products.

(5) In-depth analysis of the characteristics of different design type, such as development design, expansion design, variable parameter design, reverse design, etc., to design more pertinently according to the law.

(6) Research on the establishment of a design information base. The catalog-design information base is compiled by the system engineering method. A large amount of information needed in the design process is classified, arranged, and stored regularly, which is convenient for designers to find and call and also convenient for the computer-aided design application.

(7) Research on computer-aided design of products. The knowledge base system is established by advanced theory, and design automation is gradually realized by using intelligent means.

2) Characteristics of design methods

(1) Characteristics of traditional design methods.

The traditional design is static and mainly relies on experience and manual operation to complete. It is slow, and to a great extent, limits the design thinking progress of the human brain. 3D design objects can only be expressed by abstract 2D graphics in traditional design, and it is difficult to show the design results to engineers, technicians ,and product users. So, it is not conducive to judging, evaluating, and improving the design results.

The traditional design calculation relies on the analytic solution of engineering problems, which simplifies as much as possible, and the degree of simplification depends on whether the existing mathematical tools can be solved. On the one hand, it makes a large number of relatively complex engineering problems unable to be calculated and solved. On the other hand, it makes the solution to many engineering problems far from the actual situation.

(2) characteristics of modern design methods.

Compared with traditional design methods, modern design methods mainly have the following characteristics.

① Computerization of design means. CAD technology is widely used in the modern design process. The application of computers in design has developed from early auxiliary analysis and computer drawing to present optimal design, concurrent design, three-dimensional modeling, design process management, integration of design and manufacturing, simulation and virtual manufacturing, etc. CAD technology greatly improves design efficiency and quality,reduces the costs and lightens labor intensity. The application of network and database technology accelerates the design process, improves the design quality, facilitates the management of the design process, and facilitates information exchange among relevant departments and cooperative enterprises.

② Expansion of design category. Modern design extends the scope of design from the traditional product design to the whole product and planning process. In addition, factors such as manufacturing, maintenance, price, packaging, shipping, recycling, and quality should be considered in the design process,which is, X-oriented design.

6.1.2 Typical Modern Design Methods

With the development of the economy, science, and technology, design method develops continuously. Modern design methods and traditional design methods are only relative concepts. Modern design methods with a long history, mature theory, and wide application mainly include computer-aided design, finite element method, optimization design, and mechanical reliability design method. After the end of World War Ⅱ, developed industrial countries attached great importance to the systematic research of design theory and method. With the development of new disciplines such as design methodology, optimization design, computer-aided design, systematic design, modular design, and reverse engineering, there are many modern design methods applied in the field of engineering, such as concurrent design, similar design, robust design, green design, intelligent design, fuzzy design, virtual design, dynamic design, and so on. These new technologies are developing constantly, and they are also widely used in engineering.

Modern design methods have a wide range of contents and various disciplines. This chapter focuses on some typical modern design methods.

1) Computer-Aided Design

Computer-Aided Design (CAD) is a method and technology that uses computer hardware and software systems to assist people in designing products or projects, including design, drawing, engineering analysis, and document-making. It is a new design method and a new technology of multidisciplinary comprehensive application. CAD includes two functions: product analysis and calculation and automatic drawing. Computer, automatic plotter, and other peripheral equipment constitute the system hardware of CAD, while the operating system, file management system, language processing program, database management system, and application software constitute the system software of CAD. Generally speaking, CAD system is composed of system hardware and system software, which has the functions of calculation, graphic processing, database, etc., and can comprehensively use these functions to complete design operations. CAD is a product or process design system, which supports all stages of the design system starting from the scheme design, making the design object model, carrying out the overall design, and general layout design according to the provided design technical parameters. Through the static or dynamic performance analysis of the structure, the technical parameters are finally determined. Based on that, the detailed design and product design are completed. Therefore, the CAD system should be able to support basic design activities such as analysis, calculation, synthesis, innovation, simulation, and drawing. The basic work of CAD is to establish a product design database, graphics library, and application library.

2) Numerical calculation method

Finite Element Method (FEM) is a modern numerical method based on computer. Nowadays, this method can not only be used to solve complex nonlinear problems and non-robust problems in engineering (such as structural mechanics, fluid mechanics, heat conduction, electromagnetic field and so on) but also can be used for static and dynamic analysis of complex structures in engineering design, and can accurately calculate the stress distribution and deformation of complex parts (such as frame, steam turbine blade, gear, etc.), thus becoming a powerful analysis tool for calculating the strength and stiffness of complex parts.

The basic idea of the finite element method is as follows: Firstly, it is assumed that a continuous structure can be divided into a limited number of small meshes, which are called finite elements. The elements are only connected at a finite number of designated points, and the original structure is approximately replaced by the aggregate of constituent elements.

Equivalent node force is introduced to replace the actual vibration load acting on the element. For each unit, choose a simple function to approximately express the distribution law of element displacement components. According to the variational principle in elasticity, the relationship between node force and node displacement is established. Finally, the dynamic equation with node displacement as the basic unknown quantity can be obtained by aggregating these relationships of all elements. Given the initial conditions and boundary conditions, the dynamic equations can be solved and the dynamic characteristics of the system can be obtained. According to this idea, the calculation process of the finite element method is as follows.

(1) Structural discretization. In this stage, it will transform continuous components into multiple units.

(2) Analysis and calculation of element characteristics, that is, establishing the relationship between node displacement and node force of each element, and obtaining the stiffness matrix of each element.

(3) By using the equilibrium and boundary conditions of structural forces, the displacement of nodes and the stress values in each element can be obtained.

The idea of the finite element method is "separation and integration". First, it is for element analysis, and then it is for comprehensive analysis of the whole structure. In recent years, The application of the finite element method has developed vigorously all over the world. There are not only various general software for finite element analysis with perfect functions, such as NASTRAN, ANSYS, ASKA, SAP, etc., but also powerful pre-processing programs (automatically generating element grids and forming input data files) and post-processing programs (displaying calculation results, drawing deformation maps, isoline maps, vibration mode maps and dynamically displaying dynamic responses of structures, etc.). Because of its convenient use and high precision, the finite element program has become a reliable basis for the design and performance analysis of various industrial products.

3) Reliability design

Reliability refers to the ability of products to complete the specified functions under the specified conditions in the specified time. Reliability design is a modern design method based on probability theory and mathematical statistics, failure analysis, failure prediction, and various reliability tests aiming to ensure the product reliability. The basic process of reliability design is to select the reliability index and value of the product, distribute the reliability index reasonably, and then design the specified reliability index into the product.

4) Value Engineering

Value engineering is a design method or management science that takes function analysis as the core, development and creation as the foundation, and scientific analysis as the tool to seek the best ratio between function and cost to obtain the best value. Value engineering raises the problem that designers should stand on the user's position and produce products meeting the user's requirements at the lowest cost.

The relationship between function and cost in value engineering is:

$$V = F/C \tag{6-1}$$

Where, V——the value;

F——the function evaluation value;

C——the total cost.

Value engineering includes three basic elements: value, function, and cost.

The so-called value is the ratio between a certain function and the cost required to realize this function. In order to improve the practical value of products, we can adopt or increase the function of the products, reduce the cost of the products, or increase the function of products and reduce the cost at the same time. In a word, to improve the value of products is to make the functions of products at low cost, and the design problem of products becomes the problem of providing necessary functions to users with the lowest cost. Carrying out the research of value analysis and value engineering can achieve significant economic benefits. For example, in the 1950s and 1960s, General Electric Company of the United States spent $ 800,000 on value analysis research but gained more than $ 200 million in profits. Since China introduced value engineering technology from Japan in 1978, value engineering has also been widely used. According to statistics, the economic benefits of applying value engineering to the country have exceeded 1 billion yuan.

To improve the value of products, we can start with the following aspects.

(1) Function analysis. Start from users' needs to ensure the necessary functions of products, remove redundant functions, adjust excess functions, and increase necessary functions.

(2) Performance analysis. Study the measures to improve product performance under certain functions.

(3) Cost analysis. Analyze the cost composition and explore ways to reduce cost from various aspects.

5) Industrial design

Industrial design refers to the modeling of industrial products by artistic means in accordance with people's cognitive, psychological, and operational characteristics on the premise of ensuring the use function of products making products easy to use, aesthetic and expressive. Industrial design starts from the ever-changing needs of people, and seeks to change the way people exist with positive trends. Therefore, industrial design is not a pure art design nor is it pure plastic arts and beautiful art, but a product of the integration of science, technology, humanities, art, and economy. It develops from the comprehensive viewpoint of practicality and beauty and serves people through market communication under the constraints of science and technology, society, economy, culture, art, resources, and values.

Industrial design has the following three remarkable characteristics.

(1) Practical characteristics. Reflect the purpose of the use function, advanced, reliable

and pleasant.

(2) Scientific characteristics. Reflect the technological beauty of advanced processing means, the strict and precise beauty of large-scale industrial automatic production, and scientificity and mark the structural beauty of new achievements in mechanics, materials science, and mechanism. Under the premise of not harming the user and producer interests, we should strive to reduce the cost of products and create the highest added value.

(3) Artistic features. Apply aesthetic rules to create novel products with beauty in form, color, material, and aesthetic concept of the times embody the overall harmonious beauty of people, products, and environment. The specific contents of mechanical and electrical products modeling design are as follows: ergonomic design or agreeableness design of mechanical and electrical products, product form design, product color design, product logo, chrome brand, font design, etc.

6) Concurrent design

Concurrent design is a systematic design method that integrates product design with related processes (including the manufacturing and supporting process). In other words, concurrent design means that in the product design stage, all kinds of main performance indexes of the whole product life cycle (from concept formation to product recycling or scrap treatment) are considered at the same time to avoid unnecessary repetitive work in the later stage of product development.

Concurrent design is a research hotspot in modern mechanical design and manufacturing science. Compared with the traditional serial design method, it emphasizes that in the initial stage of product development, the influence factors of the follow-up activities of the product life cycle on the comprehensive performance of products are fully considered. The inheritance and constraint relations of performance among various stages of the product life cycle and the relations among various attributes of products are established to pursue the optimal comprehensive performance of products in the whole life cycle. With the help of the multi-functional design team composed of experts in various stages, the design process is more coordinated; the product performance is perfect so that the comprehensive requirements of users for the quality and performance of the product life cycle can be better met, and the rework in the product development process is reduced, greatly shortening the product development cycle.

The concurrent design hopes that all product development activities can be carried out in parallel in time as much as possible, which needs a higher management level to adapt to it. Concurrent design requires multi-functional teams to be closer to and understand users, more flexible and practical to develop products better meeting users' requirements. It must improve product quality, which promotes and restricts each other with the development level of design and production.

Therefore, concurrent design is a systematic and integrated modern design technology, which takes computer as its main technical means. Besides the application of CAD, CAPP,

CAM, PDM and other unit technologies in the usual sense, it also needs to focus on solving the following technical problems.

(1) Modeling and optimization of the concurrent product development process.

(2) Computer information system supporting concurrent design.

(3) Simulation technology.

(4) Comprehensive evaluation and decision-making system of product performance.

(5) Management technology in concurrent design.

7) Quick response design

Quick response design mainly includes the following contents.

(1) Establish a decision-making mechanism to quickly capture the dynamic demand information of the market.

(2) Realize the rapid design of products.

(3) Pursuing rapid trial production and finalization of new products.

(4) Implement the production system of rapid response manufacturing.

In pursuit of rapid response design of new products, we need to fully use various new technologies of manufacturing automation, such as RP (Rapid Prototyping) and VM (Virtual Manufacturing). Rapid prototyping is one of the most important developments in manufacturing technology in the past 20 years. Its characteristic of transforming CAD models into product prototypes or directly manufacturing parts at the fastest speed makes product development quickly tested, evaluated, and improved to complete design finalization or quickly form the mass production capacity of precision castings and dies.

For rapid response design, after determining product objectives, enterprises can only execute an overall design, like functional design, scheme design, and economic analysis, and then seek the best parts suppliers and manufacturers through a public information network, carry out cross-regional and cross-industry cooperation, and implement optimal combinations of production resources.

The modern design method is the application of the scientific methodology in design, and it is a new multi-cross subject. It combines the knowledge of information technology, computer technology, knowledge engineering, and management science, making the modern design method contain a wide range of contents. Because some methods still need to be perfected and developed, modern design methods can not replace traditional design methods. Some effective empirical methods are still used at present, and they are still an important component of modern design methods.

Modern design method regards the design object as a system, considers the external connection of the system at the same time, analyzes and synthesizes with the concept of system engineering, and strives for the overall optimization of the system. Modern design methods emphasize the development of creative ability, give full play to personnel creativity, and attach importance to the design, development, and innovation of the principle scheme of products.

Modern design methods emphasize integrated consideration and analysis of market demand, design, production, management, use, sales, and other factors; also, emphasis is placed on the comprehensive use of knowledge of optimization design, system engineering, reliability theory, value engineering, computer technology, and other disciplines to explore various scientific ways to solve design problems. In a word, modern design methods change empirical and analogical design viewpoints into logical, reasoning, and systematic design viewpoints and adopt dynamic, multivariable, multi-scheme, and diffusive design thinking modes. Modern design methods are systematic, creative, comprehensive, and programmatic.

6.1.3 Significance and Task of Modern Design Methods

Product design is the first step to producing products, which plays a decisive role in products performance, quality, level, and economic benefits. It has an important impact on the manufacturing process of products, as well as on the market and the whole life cycle of products. Good product design can reduce manufacturing costs, ensure product performance and service life, enhance the market competitiveness of products and produce good economic benefits. At the same time, excellent product design can also reduce the energy consumption of manufacturing and use, reduce the negative impact of manufacturing and use on the environment, facilitate the recovery and reuse of resources and benefit the sustainable development of human beings. In addition, in order to tap the market potential and open up new consumer markets, designers should invent new products or give new functions to products with innovative thinking to creat new economic growth points and enhance the competitiveness of enterprises and even a country at the background of economic globalization.

Designers are important creators of new products and have a great influence on the development of products. In order to meet the requirements of the modern science and technology, and the need of market economy system for design talents, it is necessary to strengthen the cultivation of designers' innovative ability and design quality. Designers and related engineering technicians must master modern design methods and theories skillfully and learn to use these methods and theories flexibly in practice. Only in this way can it avoid defects such as high cost, long cycle, poor performance, and large energy consumption caused by insufficient or even wrong product design, also, grasp the spark of innovative ideas in time, create new products with excellent comprehensive performance needed by the society, and continuously improve the competitiveness of enterprises. Modern design methods are primarily based on computer technology and are supported by different softwares. It is unnecessary and undesirable to memorize many formulas and deductions by rote. Therefore, the most important task of learning the modern design method is to master its basic principles and main contents, grasp the main ideas of various design methods, understand its functions and limitations, improve its design quality, and enhance its design innovation ability, and then solve practical engineering problems.

6.2 Creative Design

Innovation is an essential attribute of design. A technical scheme without any new technical elements is not a design. Only through innovative design can producers endow products with new functions, having performance beyond similar products and cost lower than similar products, having stronger market competitiveness. Today, with the high development of the knowledge economy, the life cycle of new technologies and products is getting shorter. Establishing competitive products depends on innovation, maintaining and expanding the market competitive also needs constant innovation.

Creative design means that designers adopt new technical means and principles in design, give full play to creativity, put forward new schemes, explore new design ideas, and provide designs with social value, novelty, and unique results.

6.2.1 Essence of Mechanical Creative Design

Mechanical Creative Design (MCD) refers to a kind of practical activity that gives full play to the designers creativity and makes use of the existing related scientific and technological achievements (including theories, methods, technical principles, etc.) to carry out innovative ideas and design novel, creative and practical institutions or mechanical products (devices). The mechanical innovative design includes two parts: One is improving the reliability, economy, applicability, and other technical performances of existing mechanical products in production or life. The other is creating and designing new machines and products to meet the needs of new production or life. Mechanical innovative design is a design technology and method based on the existing mechanical design theory, absorbing the beneficial components of the philosophy of science, and technology, cognitive science, thinking science, design methodology, creation science, and other related disciplines. Compared with traditional design, mechanical creative design emphasizes the dominant and creative role of people in the design process, especially in the overall scheme and structural design. Most mechanical inventions come from innovative mechanical scheme design and structural design.

Suppose engineering designers want to achieve creative design results. In that case, first of all, they must have good psychological qualities and strong dedication to work, be good at capturing and discovering the needs of society and the market, analyze contradictions, be full of imagination, and have vital insight. Secondly, they have to master creative techniques and give full play to creativity scientifically. Finally, they should be good at using their knowledge and experience, constantly improving their creativity in innovative practice, process and transforming the appearance of things on the basis of perception, and creating the thinking and imagination ability of new images.

6.2.2 Principles of Creative Design

Creative design should be carried out under some basic principles. Based on understanding the general principles of design and their interrelation and mutual restriction, attention should be paid to evaluating the design process and the final product. During the design process, designers should be good at communicating their ideas and achievements with others and improve their design ability in communication.

The basic principles of design are as follows.

(1) Innovation principle.

Innovation is the core of the design. By introducing new concepts, new methods, new ideas, new technologies, etc., designers can create objects and forms with considerable social value. The starting point of innovative design can generally be improved and broken through from the aspects of principle, structure, technology, material and process. The latest technical achievements and current design theories and methods can be used.

(2) Practical principle.

Practicality in design refers to the designed product's basic functions to achieve its purpose. It is the physical, physiological, and social functions skillfully cross-integration, full of users on the different needs of products, and high-yield product applicability. The practicality of the product comes from the design of the purpose of all people. It will change according to the crowd, the environment, and distinct personality.

(3) Economic principles.

The economic principle of design means that in the design, the product can achieve the best effect at the lowest cost. The economic principle of design usually needs to consider the cost of design, material, processing, packaging, transportation, maintenance, recycling, etc., and achieve the purpose of reducing the cost by changing the shape, reducing the volume, mass production, automatic operation and management, and adopting new technology and new materials.

(4) Aesthetic principles.

The new concept car with a novel and beautiful shape (Figure 6-1) can attract consumers to buy and obtain good economic benefits.

Figure 6-1 New concept car

(5) Moral principles.

Product design should have the moral concept of respecting other people's intellectual property rights and technological achievements. Designers must have a broad vision and a high sense of social responsibility; based on that, they can strive to design perfect products to contribute to society, maintain and improve social ethics, and promote the development of human civilization.

(6) Comply with the principles of technical specifications.

Serialization, generalization, and standardization are required in the design of mechanical products, such as bolts or standard gears used in machine tools (Figure 6-2), which can be replaced conveniently and meet the technical specification principles. The technical specifications of products are convenient for use and maintenance, which is conducive to opening up the market.

Figure 6-2 Standardized bolts, gears

(7) Principles of sustainable development.

The whole life cycle product design and the research of resource recovery and reuse should all meet the principle of sustainable development. The principle of sustainable development of technical design means that the design of products should not only meet the needs of contemporary development but also consider future development. The design of products need to concern the long-term development of human beings, rational use of resources and energy, ecological balance, and other sustainable development factors. Therefore, when designing products, it has to choose renewable resources and reusable materials as much as possible, minimize the use of raw materials and natural resources, try to reduce the loss of products in production and the energy and material consumption of products.

6.2.3 Creative Thinking Methods in the Process of Creative Design

Different designers have different creative thinking. In creative thinking, it is more important for designers to recombine and activate the information stored in their minds to form new connections. Therefore, compared with the traditional way of thinking, creative thinking shows the vitality of innovation with its breakthrough, originality, and multi-direction. According to whether the process of creative thinking follows logical rules strictly, it can be divided into intuitive thinking and logical thinking.

1) Intuitive thinking

Intuitive thinking is a thinking mode based on rich experience, reasoning, and judgment skills, which awakens and infers the problems to be solved, comprehends the essence of objects, or obtains the answers to problems. The basic characteristics of intuitive thinking are the suddenness of its production, and process and the breakthrough of its achievements. Both consciousness and subconscious play important roles in the process of intuitive thinking.

2) Logical thinking

Logical thinking is a kind of linkage reasoning that strictly follows the logical laws summarized by people based on the experience and laws of things and activities, and carries out systematic thinking. There are several ways of logical thinking, such as vertical reasoning, horizontal reasoning, and reverse reasoning.

(1) Longitudinal reasoning is to think deeply about a certain phenomenon, explore its causes and essence, and get new enlightenment.

(2) Horizontal reasoning refers to associating things whose characteristics are similar or related to a certain phenomenon and carrying out "feature transfer" to enter a new field.

(3) Reverse reasoning is to analyze the opposite aspects of a certain phenomenon, problem, or solution and find a new way.

Creative thinking is the synthesis of intuitive thinking and logical thinking. These two complex thinking processes, including gradual and sudden change, merge, supplement, and promote each other so that designers' creative thinking can be more comprehensively developed.

6.2.4 Introduction to Creative Approaches

In the actual creative design process, sometimes the inventor cannot tell what method is used to achieve success because of the complexity of the creative design process. Through the summary of practice and theory, there are roughly the following methods.

(1) Group intelligence concentration thinking.

This method of exerting collective wisdom is also known as "brainstorming". First, inform everyone of the specific functional goals. Then, after a certain amount of preparation, everyone can put forward their new concepts, methods, and ideas without any constraints and get some inspiration from others in a short time. After analysis and discussion, the false can be eliminated while the truth is preserved, from coarse to fine, and then the innovative methods and implementation schemes can be found.

(2) Bionic innovation method.

Through the analysis and analogy of biological skills in nature, the innovative design of new machines is also a standard creative design method. Due to the rapid development of the bionic method, bionic engineering has already been formed. It should be noted that there is no deliberate simulation by using this method.

(3) Reverse design innovation method.

Reverse design refers to the process of introducing advanced products from others and designing new products by analyzing and improving them.

(4) Innovative design method for analogy.

The analogy for optimization refers to the relative comparison of similar products, studying the advantages of similar products, and then designing the best products of similar products by gathering their advantages and eliminating the shortcomings.

(5) Functional design innovation method.

The functional design innovation method is a traditional design method and also a positive design method. The functional objectives are determined based on the design requirements; then, the technical scheme is drawn up and implemented, and the best design will be selected.

(6) Transplant technology innovation design method.

Transplanting technology innovation design refers to transplanting advanced technology from one field to another or applying advanced technology of one product to another to obtain new products.

(7) Computer-aided innovation method.

Mechanical innovative design is carried out using a large amount of information stored in the computer. This technology is still under development.

6.3 Virtual Technology

Globalization, networking, and virtualization have become important characteristics of the manufacturing industry, and virtual design is the essential technology of manufacturing virtualization. Virtual design is a multidisciplinary and interdisciplinary technology that relates to many disciplines and professional technologies.

Virtual reality technology provides a virtual three-dimensional environment for product creativity, change, and process optimization. In product design, designers can use this virtual environment to evaluate the virtual processing and assembly of products, thus avoiding design defects, shortening the development cycle of products effectively, and reducing development and manufacturing costs.

6.3.1 Characteristics of Virtual Technology

Virtual design is based on virtual reality technology. VR/Virtual reality technology is an advanced human-computer interaction technology developed based on three-dimensional computer graphics technology and computer hardware technology, which provide users with a realistic feeling, including three-dimensional vision, hearing, touching, and even smell and taste. Users can use natural skills, such as hand-touching, head-turning, body posture, and adjustment, to interact with the virtual world so that people become an integrated part of the system, enter immersion-interaction-conception, and virtually interact with the simulation environment built by computers. For example, enter the "virtual factory building", manipulate the "virtual machine tool", grab the "virtual parts", assemble the "virtual equipment", and so on. With the help of virtual peripherals (such as HMD, tracker, data glove, locator, etc.), people can utilize the virtual environment to complete the work that may not be realized in the real world. Virtual reality systems have the following characteristics.

(1) Autonomy.

In the virtual environment, the behavior of objects is autonomous and completed automatically by programs, which can make operators feel that all kinds of living things in the virtual environment are alive and autonomous, while all kinds of non-living things are " operable", and their behaviors conform to various physical laws.

(2) Interactivity.

Users can operate all objects (living things and non-living things) in the virtual environment, and the operation results can be accurately and truly felt by users in turn. For example, the user can directly grasp the objects in the virtual environment with his hand, get the feeling of real grasping things, and even feel the mass of the objects (in fact, there is no real object in his hand at this time). The grasped objects in the field of view move with the movement of his hand.

(3) Immersion.

In the virtual environment, users can feel different stimuli well, and the intensity of immersion is closely related to the detail, accuracy, and authenticity of the virtual environment.

The descriptive definition of virtual design can be given as follows: based on " virtual reality" technology, with the help of the mechanical product design method, designers can interact with a multi-dimensional information environment naturally through various sensors and get perceptual and rational understanding from qualitative and quantitative comprehensive integration environment to deepen concepts and germinate new ideas. Virtual reality technology has been successfully used in our daily lives, such as in the transportation sector, manufacturing sector, medical industry (artificial heart), and so on.

6.3.2 Composition of Virtual Design System

Virtual design system can be divided into two categories: enhanced visualization systems and virtual reality CAD systems.

(1) Enhanced visualization systems. The enhanced visualization system mainly uses the tradintional CAD system for modeling. Input the model built by the tradintional CAD system into the virtual environment system, and three-dimensional interactive devices (such as data gloves, three-dimensional realizers, etc.) can be used to observe the model from different angles in the virtual environment. The enhanced visualization systems usually use space balls, flying mice, etc. for navigation, and use stereo monitors with shutter eyes to enhance the

realism of the product models. At present, the enhanced visualization systems are more mature than the virtual reality CAD systems. Hence, most virtual design systems currently in use are the enhanced visualization systems. However, with the maturity of the virtual reality CAD systems, virtual design systems will gradually turn into the virtual reality CAD systems.

(2) The virtual reality CAD systems. Unlike the pure visualization system, the virtual reality CAD systems carry out three-dimensional design directly. Various input devices (data gloves, three-dimensional navigation devices, etc.) are provided to interact with the virtual environment. In addition, they can also support other input methods, such as speech recognition, gesture, and eye tracking. The virtual reality CAD systems can be mastered without systematic training. General designers can quickly get started for product design with these systems. The design efficiency of the virtual reality CAD system is 5 ~ 10 times higher than that of the current CAD system (such as Pro/Engineer).

(3) Important tools. No matter the categories of the virtual design systems, they include two necessary tools: the virtual environment generation tool and the human-computer interaction tool.

The virtual environment generation tool is the core part of the virtual design system composed of computer hardware, software development tools, and other accessories, such as sound card, graphics card, network card, and so on.

The human-computer interaction tool is the foundation to realize the "interactivity" characteristics of the virtual design system. It include helmet-mounted display, stereo earphone, tactile device, data glove, etc.

6.3.3 Features of Virtual Design

1) Virtualization as opposed to reality

The most important feature of the virtual design is virtualization. The virtual design integrates multimedia technologies such as three-dimensional dynamic display, simulation, and actual working condition simulation. Designers can feel various information such as vision, hearing, touching, and smelling and fully use their various potentials to improve the design. In scientific and technological terms, a possible feature is understood as virtual; it may appear under certain conditions. Every link of virtual design has the characteristics of virtuality and all the valuable characteristics in the real world. Although their existence is false to some extent, people can feel them and communicate in a way both sides can understand. When the tasks in the design phase are completed, some equipment (such as rapid prototyping) can transform these virtual models into real existence. Virtuality exists relative to reality, and there is a two-way mutual transformation between virtuality and reality.

2) Integration achieved by information technology

Integration is the foundation of virtual design. With the development of information technology, various intermittent and disconnected computer-aided processes can be integrated into a whole system, and the previous individual processes can be regarded as subsystems on the whole, forming a shared information flow in each subsystem. This information adapts to the data standards of different subsystems (or can be easily converted), and the content of information in circulation is also consistent. In this way, each subsystem supports the other. Subsystems often exchange and update information constantly, enriching the information base of the whole system, which is important for virtual design and the foundation for communication between virtual design and other stages of virtual product development. Integration technology ensures the internal and external information exchange of virtual design, which is beneficial for designers to adjust their idea constantly.

3) Dynamic human-computer interaction

Virtual design is a process of dynamic thinking and operation. Using various interactive means (data gloves, sounds, commands, etc.) to support more design behaviors (modeling, simulation, prediction, evaluation, etc.), designers can modify the virtual model, and the virtual model will immediately respond accordingly, and then the designers can see their modified results. This human-computer interaction process helps designers to fully express their multiple ideas and make the virtual model more detailed. Using computer modeling, it can place the virtual camera conveniently in any position, either on the usual visual horizon or in a position that cannot be reached in the real world. From these points of view, people can see the scene effect that they are used to and can also see unimaginable scenes. With the development of multimedia technology, people can observe the model from static to dynamic and watch smooth model animation. Moreover, multimedia technology can stimulate other sensory systems of people. People appreciate the model itself deeply understand the creativity of design, and even enjoy the design and production process.

4) Digitalization of information interaction

The information of virtual design is stored digitally, and digitalization is the critical part to information flow in virtual design. Virtual design data has relevance; therefore, the modification of data for each subsystem will immediately affect the whole virtual design process, and every related data will make corresponding responses, which lets designers know the impact of modifications on the overall situation quickly. Another advantage of digitalization is that it can be stored and called conveniently, and the whole history of the design process can be recorded. It enables the designer to make adjustments according to the history of the design process, such as Model Tree in Pro/E. Good modifiability ensures the efficiency of product design and development.

The visualization provided by virtual technology is not only the spatial display of general geometric shapes but also the visualization of noise, temperature change, force change, wear and vibration, etc. It can also express people's innovative thinking as visual virtual entities and promote people's creative inspiration to further sublimate. Therefore, in the virtual state, it can visually track and describe the whole process of the product life cycle (design, processing and manufacturing, assembly, performance analysis, use and recycling, etc.). More emphasis is placed on product design's high feasibility and reliability before the product is processed to minimize the risk of product development in the subsequent stages, such as manufacturing with a large proportion of investment funds.

The main advantages of developing products with virtual technology are as follows: It can promote creative design, accelerate the product development cycle, reduce waste of resources, reduce the risk of product development, promote remote collaborative product development, improve the quality of technical training and education and enhance the competitiveness of enterprises.

5) Features of real product and virtual product development

Figure 6-3 and Figure 6-4 show the real product development process and the virtual product development process.

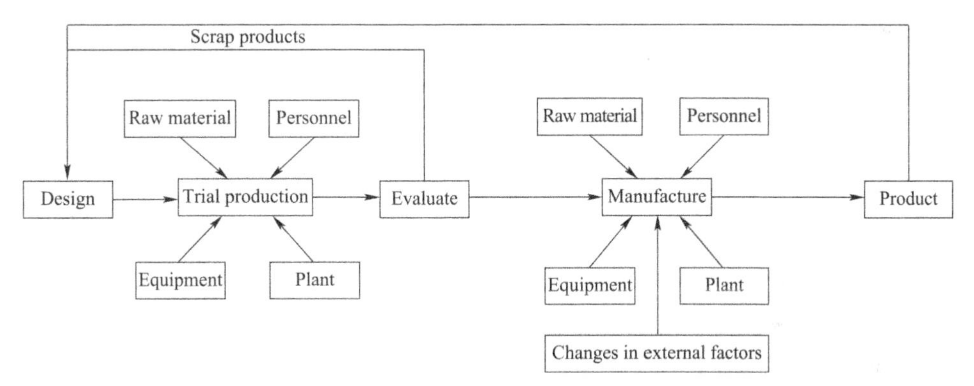

Figure 6-3 Realistic product development processes

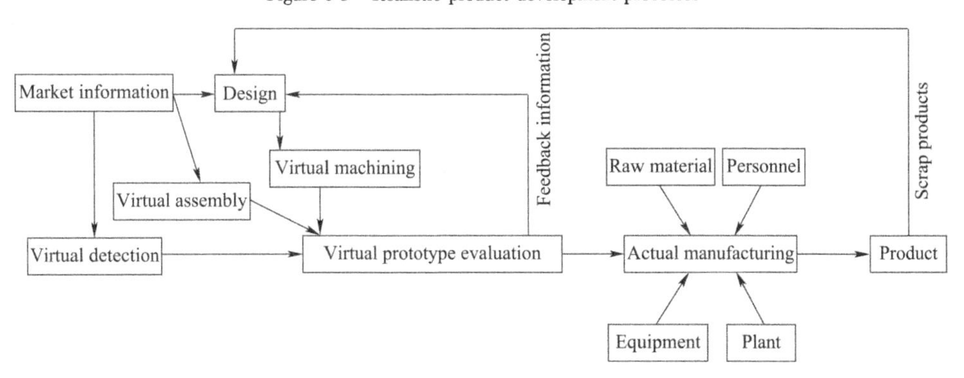

Figure 6-4 Virtual product development process

At present, the revolutionary influence of virtual design on traditional design. The product design, development, and processing are all done on computers for the virtual design system. Therefore, it does not consume resources and energy or produce actual products. Compared with traditional design and manufacturing, virtual design has the characteristics of high integration, rapid prototyping, distributed cooperation, and so on. The specific advantages are as follows.

(1) Virtual design inherits all the characteristics of virtual reality technology.

(2) Virtual design inherits the advantages of the traditional CAD design and is convenient to use the original achievements.

(3) Virtual design has simulation technology's visualization characteristics, making it convenient to improve and revise the original design.

(4) Virtual design supports collaborative work and design in different places, which is conducive to resource sharing and complementary advantages. It is convenient to use and supplement various advanced technologies and maintain the leading edge in technology.

6.3.4 Critical Technologies of Virtual Design

(1) Real-time dynamic display technology of three-dimensional images.

Three-dimensional vision is a vital information feedback channel of the virtual design system. Therefore, it is required to use fast processing methods in image modeling and stereo image generation to achieve the best real-time display of three-dimensional stereo images.

(2) Sound system in the virtual environment.

The auditory channel is one of the most important interfaces in the virtual environment; and it is the second information channel after visual feedback. It mainly involves three-dimensional virtual sound modeling and three-dimensional virtual sound system reconstruction.

(3) Contact feedback and strength feedback.

Tactility is an essential part of the virtual design. If people can operate the objects in the virtual environment with their own hands and get rich sensory information, they will greatly enhance the immersion and realism of the virtual environment, which improves the accuracy and efficiency of the performing tasks.

In a word, virtual design based on virtual reality technology will help designers improve product quality, shorten the product development cycle, and reduce development costs.

6.3.5 Development Trend of Virtual Technology

(1) A new VR-CAD system.

Highly interactive, immersive, and three-dimensional design environment can carry out virtual surface modeling and drag the control vertices of the surface in three-dimensional space. Virtual sculpture modeling, "virtual hand" modifies and manipulates the surface geometry of three-dimensional objects.

(2) Collaborative virtual design.

Immersive (conducive to the actual scene) and semi-immersive (conducive to communication between users) collaborative virtual design is conducive to optimal design.

The era of the knowledge economy provides an opportunity to innovative design ideas and methods. It should be fully utilized modern and high-tech innovative design methods and technologies to update traditional product design methods, improve design efficiency and quality, and develop more products with market competitiveness and independent intellectual property rights. Virtual design technology, not only in the scientific and technological community but also in the business community, has caused widespread concern and has become a research hotspot.

6.4 Optimization Design

The term optimization is often used to mean "improvement", but mathematically, it is a much more precise concept: finding the best possible solution by changing variables that can be controlled, often subject to constraints. Optimization has a broad appeal because it is applicable in all domains and because of the human desire to make things better. Any problem where a decision needs to be made can be cast as an optimization problem. Optimization is often used during engineering design process. It is a systematic process that uses design constraints and criteria to find an optimal solution. A wide range of optimization techniques and methods is available for researchers and designers, and they are selected in accordance with the nature of the optimization tasks, applications involved, and designer's expertise.

6.4.1 Basic Concept of Optimization Design

Optimization design is a new subject developed at the beginning of the 1960s. It is gradually formed by the development of optimization theory and computer technology and provides an essential scientific method for engineering design. Using this method, designers can find the best design scheme, thus greatly improving the design efficiency and quality. Therefore, optimization design has become an important field of modern design theory and method, which is widely used in various industrial departments and plays an important role in improving product performance, product quality, and design efficiency.

Applying the optimization method in mechanical design can not only make the scheme achieve some optimization results under the specified design requirements but also avoid too much calculation workload. The reasons which mechanical optimization design has been widely used in mechanism synthesis, general parts design of machinery, various special machinery design, and process design are as follows: On the one hand, there are a lot of optimization design problems in production and engineering design that need to be solved urgently. On the other hand, the development of computer technology and the popularization of scientific computing software provide an effective tool for adopting optimization technology. The optimization of product structure and production process has become an effective means of market competition.

Mechanical optimization design is a process of finding a set of design parameters to satisfy certain constraints so that single or multiple design indexes of mechanical products can reach the optimum. It mainly includes two aspects: Establishing the mathematical model for the optimization design problem and selecting the appropriate optimization method and program. The mathematical model that describes the actual design problem in mathematical form should be established first in terms of the actual mechanical design problem. It is necessary to apply professional knowledge to determine design's limiting conditions and goals and establish the interaction between design variables. The model can be in different forms, such as analytical formula, experimental data, or empirical expression, reflecting the quantitative relationship between design variables. Once the mathematical model is established, the optimization design problem becomes a mathematical solution problem. It can choose or make programs based on the characteristics of the mathematical model to get the best design parameters.

MathWorks Company in the United States launched the scientific computing software MATLAB in 1994. It is powerful in scientific calculation, graphics processing, and data visualization. The attached Optimization Toolbox contains a series of optimization algorithms and modules that can be used to solve large-scale constrained optimization problems. In addition, there are optimization design modules in ANSYS, an international general large-scale finite element software, and ADAMS, a digital functional prototype software. The popularization and application of the software mentioned above provide an excellent platform for engineers and technicians to solve optimization problems.

6.4.2 Mathematical Model of Optimization Design

To use the optimization design method for mechanical design, we must first express the design problems in mathematical language and establish a mathematical model. The mathematical model of optimal design has two parts: design goal and design constraint. The model should be precise and concise in abstracting design problems. "Precise" means that the model cannot be distorted, which is the most basic requirement for the mathematical model. "Concise" means that the model should not be too complicated so as not to bring unnecessary difficulties to the solution. The mathematical model of optimal design includes the following three elements.

(1) Design variable.

A set of values of basic parameters can express a design scheme. These basic parameters can be the structural dimensions and position relationships of the design objects. It may also be physical quantities such as the elastic modulus and allowable stresses of the material, the weight and speed of the parts, etc. It can also be an output of deformation, natural frequency, efficiency, etc., which represents the working performance. The parameters that can be pre-determined according to existing experience are design parameters, such as some parameters of working performance, process, and structure arrangement, etc. Other basic parameters that need to be modified and adjusted constantly in the optimization design process are called design variables or optimization parameters, and their changes will directly or indirectly affect the value of the objective function.

The entire design variable is a set of variables that a column vector can represent:

$$x = [x_1, x_2, \cdots, x_n]^T \tag{6-2}$$

Where, x——the design variable vector.

The order of components in a vector is completely arbitrary and can be chosen arbitrarily according to the convenience of use. Once such a vector is specified, any particular vector can be considered as a "design". The real space consisting of N design variables in coordinates is called design space.

(2) Objective function.

The objective function is a quantitative indicator for the designer to measure the performance of the optimization scheme and is an important issue in the whole optimization design process. Some designs are better than others in all possible designs, and all those designs must have some better properties. If this property can be expressed as a calculable function of the design variable, then optimization of this function can be considered for a better design. The function used to optimize the design is called the objective function. Sometimes, it is used to evaluate the design scheme, which also called the evaluation function and recorded as $f(x)$.

Many design problems may have more than one design objective. The problem with only one design objective is called a single objective optimization problem, while the problem with more than one design objective is called a multi-objective optimization problem. The solution of multi-objective optimization is very complex, and the method is also not as mature as the single-objective. If one design objective is much more important than the other, it can be considered as the design objective for a single-objective optimization problem, and the other design objectives can be treated as constraints in the mathematical model.

(3) Constraint condition.

A design that meets all the requirements is called a viable design, and vice versa is called an infeasible design. A viable design scheme must satisfy certain design constraints, called restraint conditions or restraints.

Constraint conditions can be divided into performance constraints and boundary constraints. Performance constraints are from the performance requirements, including strength conditions, stiffness conditions, vibration stability conditions for part design, and assembly conditions, speed conditions, transmission ratio conditions for mechanism design, etc. Many performance constraints are calculation formulas of design specifications that are derived from knowledge of mechanics, and geometry. Boundary constraints are constraints on the range of design variables and give the boundaries of design variables. Boundary constraints are given based on the structural needs or experience of the design object, and neither the necessary boundary constraints can be omitted, nor the boundaries can be narrowed without basis.

Constraints can also be divided into equality constraints and inequality constraints according to their mathematical expressions.

For the equality constraints:

$$h(\boldsymbol{x})=0 \tag{6-3}$$

where, $h(\boldsymbol{x})$——equality constraints.

The design point is required on the constrained surface of the N-dimensional design space. For the inequality constraints:

$$g(\boldsymbol{x})\leqslant 0 \tag{6-4}$$

where, $g(\boldsymbol{x})$——inequality constraints.

The design point is required to restrain the side of the surface $g(x)=0$ (including the surface itself) in the design space. Therefore, constraints impose limits on the scope of activity of the design point in the design space.

In optimum design, only the design scheme that satisfies all the constraints is available. Therefore, the necessary design constraints must be listed in the mathematical model to ensure no omissions since omitting important constraints could lead to the wrong design scheme. In addition, it should also pay attention to the fact that the constraints cannot contradict each other; otherwise, no results will be obtained.

6.4.3 General Form of Mathematical Model of Optimization Design

The mathematical model of the optimization problem is a mathematical abstraction of the actual optimization design problem. After defining the design variables, constraints, and objective functions, the optimization design problem can be expressed in a general mathematical form.

Finding design variable vectors:

$$\boldsymbol{x}=[x_1,x_2,\cdots,x_n]^T \tag{6-5}$$

Make

$$f(\boldsymbol{x})\rightarrow \min \tag{6-6}$$

And satisfy the constraints:

$$\begin{cases}(h_k(\boldsymbol{x})=0, & k=1,2,\cdots,l\\ g_j(\boldsymbol{x})\leqslant 0, & j=1,2,\cdots,m\end{cases} \tag{6-7}$$

Where, k——the *kth* equality constraints;

g_j——the *jth* inequality constraints;

l——the number of the equality constraints;

m——the number of the inequality constraints.

In practical optimization problems, there are two types of requirements for objective function: minimization $f(x)\rightarrow\min$ or maximization $f(x)\rightarrow\max$. Since the maximization of $f(x)$ is equivalent to the minimization of $f(x)$, the mathematical expression of the optimization problem in this book is always objective function minimization.

According to the existence of design constraints in equation (6-7), optimization problems can be divided into constrained and unconstrained optimization problems. When the mathematical model does not contain any constraints, it is called an unconstrained optimization problem; otherwise, it is called a constrained optimization problem. Constrained optimization is usually the actual optimization design problem. Since constrained optimization problems can be solved as a series of unconstrained optimization problems, the solution of unconstrained optimization problems is an important research content in optimization design.

According to the properties of the objective function and constraint function in equation (6-7), optimization problems can be divided into linear and non-linear planning problems. When the objective function and all the constraint functions are linear functions of the design variables, it is called a linear planning problem; otherwise, it is called a non-linear planning problem. Generally, many optimization problems in production plan management are linear planning problems. Most optimization problems in engineering design are non-linear planning problems.

6.5 Green Design

Green design is a new design concept and method put forward in the early 1990s on the subject of how to save resources, utilize energy, and protect the environment effectively while developing the economy. It is one of the effective ways to show sustainable development and has become a hot spot and main content of modern design technology research.

6.5.1 Background and Concept of Green Design

The research shows that the damage to the ecological environment caused by product design is much greater than that caused by the design itself.

In traditional design, the designer usually considers the function, quality, life, and cost of the products. The design principle is the product should be easy to manufacture and meet the required functions and performance. However, little or no consideration has been given to the recycling of resources and the impact of products on the ecological environment. It will lead to low recovery and utilization and the serious waste of resources and energy, especially toxic and harmful substances, which will seriously pollute the ecological environment and affect the sustainability of production and development. Therefore, the product should be planned and designed according to the characteristics of green products at the beginning of the design. Only in this way can products' final "green" characteristics be guaranteed. At present, industrially developed countries strive to achieve miniaturization (less material), multi-function (multi-use, less land occupancy), recyclability (reduce the waste quantity and pollution) in product design and pursue energy-saving, material-saving, waste-free, closed-circuit cycle in production technology. All of these are effective means to achieve the green design. However, green design involves many disciplines (e. g., mechanical discipline, material discipline, environmental protection, social science, and management discipline, etc.), covers a wide range, belongs to the interdisciplinary research field, and is still in continuous improvement. Therefore, implementating green design also needs integrated analysis and coordination from a systematic point of view, considering both technology and management aspects, and requires joint efforts from the whole society.

Green design is also called ecological design and environmental design. Although the designations are different, the connotations are consistent. The basic idea is to incorporate environmental factors and pollution prevention measures into product design, take environmental performance as the design objective, and attemp to minimize the impact of products on the environment. It is necessary to reduce not only the consumption of substances and energy but also the discharge of harmful substances and to facilitate the classification and recycling of products and parts.

In summary, green design is a design that focuses on the environmental attributes of products (disassembly, recyclability, maintainability, reusability, etc.) throughout the product life cycle and makes it a design goal. Meanwhile, the basic functions, service life, and quality of the product are guaranteed. It requires that reasonable raw materials, structures, and processes must be used in the design of products by environmental protection indicators, energy consumption is reduced, no toxic side effects are produced in the manufacturing and use process, the products are easy to disassemble and recycle, and the recycled materials can be used for reproduction.

6.5.2 Green Products and Green Design Principle

1) Green Products

Green design is a design method developed for green products, so before introducing green design, it need to be understood what green products are. Green Product (GP) or Environmental Conscious Product (ECP), compared with traditional products, has not yet been recognized as an authoritative definition due to the unclear description and quantitative characteristics of its "green degree". However, by analyzing the existing different definitions, we can still get a basic definition of green products. Green products refer to those that are harmless to or have little harm to the ecological environment, have a high resource utilization rate, have the lowest energy consumption during the whole life cycle of the products, and meet the environmental protection requirements of their characteristics. Therefore, green products have rich connotations, which are mainly manifested in the following aspects.

(1) Excellent environmental friendliness.

All product links from production to use, even waste, recycling, treatment, and disposal, are not harmful to the environment or do little harm. It requires enterprises to select clean raw materials and clean processes in production, little or no environmental pollution occurs and no harm to the user when using the product, rarely generating waste for the scrapped products during recycling.

(2) Maximize the material resources utilization.

Green products should minimize the materials consumption and reduce the types of materials used, especially rare and expensive materials, as well as toxic and harmful materials. This requires simplifying the product structure as much as possible when satisfying the basic functions of products, selecting materials reasonably, and reusing the parts and materials as much as possible.

(3) Maximize energy conservation.

Green products should consume the least energy at all stages in their life cycle. The economical use of resources and energy is also a good means of environmental protection.

2) The principle of green design

Green design ought to be based on the principle of green technology. The so-called green technology is a general term for the technology, process, or product that reduces environmental pollution or the use of raw materials and natural resources. The purpose of green design is to overcome the shortcomings of traditional design so that the designed products can meet the requirements of green products. It contains the concept of fundamentally preventing pollution, saving resources and energy, covers from formation, manufacturing, use, disposal of recycling, reuse, and treatment, etc. , the entire life cycle of the product. The basic idea of green design lies in preventing the harmful effects of products and processes on the environment in advance and then manufacturing.

At the stage of conceptual design and rough design, considering fully all kinds of environmental impacts after the product is manufactured, sold, used, and scrapped, the technicians should work closely together, share information, and use environmental assessment criteria to restrict the design processes of manufacturing, assembly, disassembly, and recycling, and make them economically sound. Green design must follow certain systematic design procedures, including assessment of environmental regulations, identification of environmental pollution, raising environmental problems, reducing pollution, alternatives meeting user requirements, technical and commercial evaluation of alternatives, etc. Green designers should consider such issues as what waste may be generated during manufacturing, what substitutes toxic ingredients are, how waste products are managed, what impact design has on product recycling, what impact parts and materials have on the environment, and how users use products.

Green design usually has three main stages.

(1) Track materials and balance between the input and the output of materials.

(2) Allocate environmental costs to special products or product categories and determine the corresponding product value.

(3) Systematically study the design process rather than focusing on the product itself. Considering the overall quality of products, designers should not only design products according to physical objectives but also take the services provided by products as the primary basis.

6.5.3 Main Contents and Key Technologies of Green Design

Green design involves many disciplines, such as mechanical manufacturing, materials science, management, sociology, and environmental science, with strong interdisciplinary characteristics. Obviously, it is difficult to adapt to the requirements of green design only by relying on one of the existing design methods. Therefore, green design is a developing system design method that integrates object-oriented technology, concurrent engineering, and life cycle design. The design system integrates product quality, function, life, and environment. It is the concrete embodiment of the idea of sustainable development. Figure 6-5 shows the outline of the green design. Green design should focus on the whole product life cycle process, not on a certain stage, link, or department.

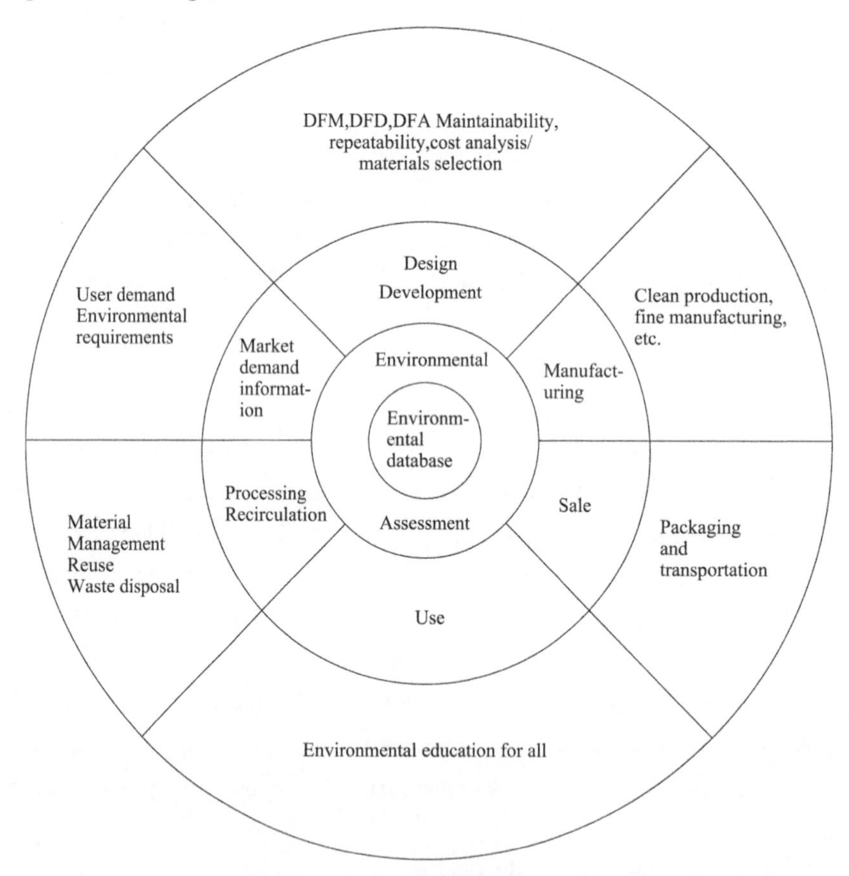

Figure 6-5 Process outline for green design

1) Main contents of green design

(1) Description and modeling of green products.

The keypoint to green design is to describe green products accurately and comprehensively and establish a systematic evaluation model. For example, for the products that have been produced, the evaluation index system and the principles for formulating evaluation standards have been put forward, the "green degree" of the products has been evaluated using the fuzzy evaluation method, and the corresponding evaluation tools have been developed.

(2) Material selection and management of green design.

Green design requires product designers to change the traditional material selection procedures and steps. While selecting materials, they should consider not only the use conditions and performance of products but also the environmental constraints. At the same time, they must understand the impact of materials on the environment and select non-toxic, pollution-free materials that are easily recycled, reused, and degraded. The requirements of green design for materials also pose new challenges for the development of materials science. Moreover, material management should also be strengthened. The material management of green product design includes two aspects: On the one hand, materials containing harmful ingredients and harmless ingredients cannot be mixed. On the other hand, the valuable parts of the products shall be fully recycled, and the parts that cannot be used shall be treated by certain processes to minimize their impact on the environment.

2) Product recyclability design

Recyclability design considers a series of issues related to recyclability, such as the possibility of recycling, the value of recycling, the method of recycling treatment, and the processability of recycling treatment structure at the initial stage of product design, to achieve the maximum utilization of material resources and energy, and minimize environmental pollution.

The main contents of recyclability design include: ①recyclable materials and their marks; ②recyclable processes and methods; ③ economic evaluation of recyclability; ④ recyclable structure design.

3) Removable design of products

Modern mechanical and electrical products often use a variety of different materials, so their disassembly has become the main focus of green design research. Non-removable will not only cause a lot of reusable parts and materials waste but also seriously pollute the environment on account of the difficult disposal of waste. Disassembly plays an important role in the benign development of modern production and has become an important part of mechanical design. Removable is one of the main contents of green product design. It should be considered in the initial stage of product design, for instance, the designed structure should be easily disassembled and maintained, and the reusable parts should be effectively recycled and reused after the product is scrapped. It requires changing the product structure design's traditional connection mode into a detachable one. There are several ways to do it: One is the "case" method based on mature structures, and the other is the automatic design method based on computers. A systematic evaluation index system, evaluation methods, and design criteria for disassembly structures have been proposed, and corresponding evaluation software has been developed.

4) Cost analysis of green products

The cost analysis of green products is different from the traditional cost analysis. At the initial stage of product design, it is necessary to consider product recovery, reuse, and other performance. Therefore, when analyzing the cost, it is necessary to consider the replacement of pollutants, product disassembly, reuse, and the corresponding environmental costs of particular products. Paying for environmental protection for enterprises will also lead to different product costs. The actual costs of the same environmental projects in different countries or regions will also result in different costs among enterprises. Therefore, the cost analysis of green products should be carried out in every design decision to make sure that the designed products have a high "green degree" and low overall cost.

5) Green Design Database

The green design database is enormous and complex. The database plays an important role in the design process of green products. It shall include all data related to the environment and economy in the product life cycle, such as material composition, the impact of materials on the environment, natural degradation cycle of materials, artificial degradation time, cost, the additives generated during manufacturing, assembly, sales, and use, as well as various judgment standards required by environmental assessment criteria.

6) Key technologies of green design

It can be seen that the key technologies of green design include the following aspects.

(1) Research on the evaluation system and method of green products.

(2) Establishment of the green design model.

(3) Collection and arrangement of green design data and establishment of the database.

(4) The system research of green design method and the establishment of the knowledge base.

6.5.4 Developing Trends in Green Design

Product combination design, cycle design, and dematerialization design of products and services have become a trend of "green design", which can better reflect energy conservation and environmental protection. The theme and development trend of "green design" is roughly reflected by the following three aspects.

(1) Use natural materials.

It is used in furniture, building materials, and fabrics as "unprocessed". The design of direct use of natural materials in products not only saves energy but also shortens production links and improves production efficiency.

(2) Emphasize the economy of materials.

Discard useless functions and purely decorative styles and return to classic simplicity. At the same time, the factors of "high-tech" and "high emotion" are carefully integrated into the simplicity. It makes people feel fashionable and close when using the product.

(3) Use recycled materials.

Recycling is one of the core strategies that can enhance the aesthetic and functional value of green design, by saving resources, reducing waste, and creating unique design features. Recycled materials can come from various sources, such as demolition waste, industrial by-products, or household items. For example, it can use reclaimed wood, metal, or bricks for structural or finishing elements, or repurpose glass bottles, plastic containers, or tires for creative items. Recycled materials can reduce the need for virgin materials, which saves energy, water, and land resources, and realizes lowers green house gas emissions.

REFERENCES

[1] 崔玉洁,石璞,化建宁. 机械工程导论[M]. 北京:清华大学出版社,2013.
[2] 郭绍义. 机械工程概论[M]. 武汉:华中科技大学出版社,2009.
[3] 张宪民,陈忠. 机械工程概论[M]. 武汉:华中科技大学出版社,2011.
[4] 邹慧君,高峰. 现代机构学进展[M]. 北京:高等教育出版社,2007.
[5] 黄宝强. 走进科学与技术[M]. 上海:复旦大学出版社,2004.
[6] 王中发,殷耀华. 机械[M]. 北京:新时代出版社,2002.
[7] 李杞仪,李虹. 机械工程基础[M]. 北京:中国轻工业出版社,2010.
[8] 陈永久. 机械基础[M]. 长沙:国防科技大学出版社,2006.
[9] 李光布,饶锡新. 机械工程专业英语[M]. 武汉:华中科技大学出版社,2008.
[10] 孙大涌. 先进制造技术[M]. 北京:机械工业出版社,2002.
[11] 张世昌. 先进制造技术[M]. 天津:天津大学出版社,2004.
[12] 楼锡银. 机电产品绿色设计技术与评价[M]. 杭州:浙江大学出版社,2010.
[13] 刘飞. 绿色制造的理论与技术[M]. 北京:科学出版社,2005.
[14] 李佳. 计算机辅助设计与制造[M]. 天津:天津大学出版社,2003.
[15] 艾兴. 高速切削加工技术[M]. 北京:国防工业出版社,2003.
[16] 张伯霖. 高速切削技术及应用[M]. 北京:机械工业出版社,2002.
[17] 王先逵. 精密加工和纳米加工 高速切削 难加工材料的切削加工[M]. 北京:机械工业出版社,2008.
[18] 王广春,赵国群. 快速成形与快速模具制造技术及其应用[M]. 北京:机械工业出版社,2008.
[19] 刘光富. 快速成形与快速制模技术[M]. 上海:同济大学出版社,2004.
[20] 吕仲文. 机械创新设计[M]. 北京:机械工业出版社,2004.
[21] 梅顺齐,何雪明. 现代设计方法[M]. 武汉:华中科技大学出版社,2009.
[22] 吕宏,王慧. 机械设计[M]. 北京:北京大学出版社,2009.
[23] 仝勖峰. 机械工程概论[M]. 北京:电子工业出版社,2015.
[24] BIRD J, ROSS C. Mechanical engineering principles[M]. London; New York: Routledge, 2014.
[25] SHIGLEY J E, MISCHKE C R, Brown Jr T H. Standard handbook of machine design[M]. New York: McGraw-Hill Education, 2004.
[26] CLIFFORD M, SIMMONS K, SHIPWAY P. An introduction to mechanical engineering: Part 1[M]. London: CRC Press, 2009.
[27] DARBYSHIRE A. Mechanical engineering[M]. London; New York: Routledge, 2011.
[28] ATKINS A G, ESCUDIER M. A dictionary of mechanical engineering[M]. Oxford: Oxford Quick Reference, 2013.